A
Brief History of Light
and Those That Lit the Way

A
Brief History of Light
and Those That Lit the Way

Richard J Weiss

World Scientific
Singapore • New Jersey • London • Hong Kong

Published by

World Scientific Publishing Co. Pte. Ltd.

P O Box 128, Farrer Road, Singapore 912805

USA office: Suite 1B, 1060 Main Street, River Edge, NJ 07661

UK office: 57 Shelton Street, Covent Garden, London WC2H 9HE

Library of Congress Cataloging-in-Publication Data
Weiss, Richard J. (Richard Jerome), 1923–
   A brief history of light and those that lit the way/by Richard J. Weiss.
      p.   cm.
   ISBN 9810223773. -- ISBN 9810223781 (pbk.)
   1. Light.   2. Optics.   I. Title.
QC355.2.W43   1995
 535'.09--dc20   96-21374
                                        CIP

British Library Cataloguing-in-Publication Data
A catalogue record for this book is available from the British Library.

Cover illustration M. Behr 1990
   Artistic rendering of the Eddington Experiment (1919) depicting the gravitational attraction of starlight by the sun during a solar eclipse.

Editorial Layout, Conroy Concepts (Donna J. Conroy) Brockton, MA

Printed in Singapore.

Contents

A BRIEF HISTORY OF LIGHT
AND THOSE THAT LIT THE WAY

Cover Illustration: *M. Behr*

Preface

Since the dawn of life on earth, light has shaped man's understanding of the natural world and controlled his movements over land and sea. From the terror engendered by jagged flashes of lightning to the awe-inspiring beauty of a rainbow, light in its infinite manifestations has surrounded people and demanded attention. Terror and wonder joined to produce ancient mythologies, religious beliefs, rites of human behavior, artistic representation, and - ultimately - scientific curiosity.

That the universe could be understood through the twin agencies of observation and reason constituted the essence of the Greek miracle and the foundation for scientific inquiry. Terror and awe were gradually displaced by increased understanding. Ancient and medieval learning became grist for the Renaissance and the scientific Revolution of the seventeenth century. Observation of terrestrial and celestial phenomena was expanded enormously through the invention of new instruments, and reasoning about nature was guided and constrained by powerful new mathematical techniques.

And everywhere was light. It passed through Galileo's telescope from the moons of Jupiter, through Grimaldi's narrow slits to a nearby screen, through Hooke's microscope from the wings of a fly, through Huygen's crystal of Iceland spar to a flat surface, and through Newton's prism to a distant wall. Light was refracted by the earth's atmosphere and by lenses, and doubly refracted by certain curious crystals. It was reflected by water and glass surfaces. It was diffracted by sharp edges and narrow slits. It was dispersed into a spectrum of colors by raindrops and prisms. And its behavior in single and double refraction, reflection, diffraction, and dispersion could be described by mathematical laws.

But the deepest mystery of all remained unfathomed and subject to intense dispute. What was the ultimate nature of light? Hooke and Huygens argued for waves or pulses, Newton for particles. Continuous motion or discrete substance?

Two centuries elapsed before the veil of this profound mystery was lifted. The wave-particle dispute continued and increased in intensity in the early decades of the nineteenth century, with Young and Fresnel foremost on the wave side of the question, Laplace and Biot on the particle side. Increased mathematical sophistication, especially the differential and integral calculus, was brought to bear, and new and more precise instruments of observation - even by mid-century ones for measuring the incredibly high velocity of light - added insight. Conviction was attained at last in the final decades of the century through Maxwell's magisterial synthesis, his electromagnetic wave theory of light, and its experimental confirmation by Hertz.

But conviction is not synonymous with reality. Einstein reopened the question anew in a revolutionary way in 1905 by the introduction of his light-quantum hypothesis, which defined conclusive verification for almost two decades, until the experiments of Compton. Soon thereafter, through the deep analyses of Heisenberg and Bohr, the ancient dichotomy of wave *or* particle was replaced in the new quantum mechanics by the inclusive complementarity of wave *and* particle, both aspects being necessary for a complete understanding of the nature of light.

The recorded history of light thus spans millennia, and represents one of the greatest triumphs of the human intellect. But no one should conclude that the story is over. The quest for scientific understanding is never ending, because scientific curiosity is inextricably rooted in human nature and cannot be stifled. Moreover, while some scientists make extraordinary contributions and hence garner much attention over time, countless others invent new instruments, gather additional experimental data, and provide important theoretical insights that greatly broaden the foundation for scientific achievement and its exploitation for cultural enhancement and technological gain. And it must never be forgotten that each scientist is a unique human being, with unique talents, skills, wants, desires, and foibles. The pursuit of science is as intricate and complex as the scientists themselves and the diverse societies in which they live and work. Science is a profoundly human and social activity.

The history of science in general, and the history of light in particular, is therefore replete with surprises. Who might have expected, for example, that the great Rembrandt would achieve textures of light and color that would inspire scientific and psychological inquiry, that the thoroughly unscrupulous opportunist Rumford would become a splendid scientist and patron of science, that the brilliant scientific work of Young would be met with ridicule, that the inventor Fox Talbot would fail to get his due because of judicial incompetence, that the colonial farm boy Rutherford would become the greatest experimental physicist of his day, or that the purely theoretical analysis of Einstein would be essential for the invention of the laser a half-century later? Only the ever-changing interplay of complex psychological, educational, cultural, and social forces could produce results such as these.

This book traces the interplay of these forces in the history of light from the seventeenth century to the present. It provides engaging portraits of many of the scientists whose achievements constituted that history, and of major inventors and artistic and literary figures who contributed to or were influenced by it. And along the way it provides tutorials designed to facilitate understanding of concepts potentially unfamiliar to some readers. The result is a book written in a style eminently suitable for a general audience of readers.

Roger H. Stuewer
University of Minnesota
March 16, 1995

A BRIEF HISTORY OF LIGHT AND THOSE WHO LIT THE WAY

Introduction

The combination of the human eye as a detector and the brain as an information processor has achieved a sophistication unmatched in our solar system, or by any computer. Take, for example, the ability to recognize within a fraction of a second any one of thousands of faces and places, even after an absence of years. There is more to this working relationship of sight and brain than meets the eye for psychological factors also come into play. The images we see can instantaneously arouse a myriad of emotional responses from fear to friendship to love. The eye, and its visible light receptors, is our primary tool to cope in our world. Seeing is believing.

Man has been clever enough not to limit himself to the visible but has harnessed the 'invisible' portion of the electromagnetic spectrum from radio waves to gamma rays. Yet, he still feels most comfortable to have these signals converted into visible images such as in viewing TV or in analyzing X-rays through radiographic film. By storing information through our visual memory bank we achieve the highest confidence level in dealing with our environment.

An appreciation for the history of light not only has to direct its attention to what we see, but to how it affects what we feel and do. Man has studied light as a physics problem and has manipulated light in art and cinema to arouse a multitude of responses, yet he is still in the dark as to how the brain makes the most effective use of the light signals received by the eye.

This is the brief story, from Leonardo to Oppenheimer, of what we know about light as we enter the last decade of 500 light years of exploration, from the perception of the subject during the Renaissance to the revolution spawned by the laser and optical fibers.

Richard Weiss
King's College London

CHAPTER I

16th and 17th Centuries

God said 'Let There Be Light!'

And Einstein planted a seed

Maiman created the laser one night

And sated man's optical need!

The laser, theorized by Einstein early in the century and fabricated by Ted Maiman in 1960, may not be the Messiah but its impact on society may be equally astounding. We are able to clone light particles, called photons, into battalions of indistinguishable soldiers all marching in precise step and guided along optic fibers to do multifarious tasks from illumination and surgery to cutting holes in diamonds. Never before has man been granted such minute and exact control over the Creator's fundamental building blocks. And this gift appears to be the exclusive province of man; no laser light reaches us from outer space! Copernicus and Galileo overturned Ptolemy's description of the world as the center of the universe - the laser may again cede to homo sapiens a 'geocentric' focus for the planet earth.

Light as an entity, to be examined for itself, began seriously with men like Leonardo. For millennia it was taken for granted although sun worshipers date back to the Stonehenge days. Leonardo employed light and his artistic creativity to produce the most famous painting in the world. He fabricated various devices to focus and control light rays and he ranks with Einstein in his power to question the fundamentals of its physics. Einstein's inquisitiveness was spurred on by the undetectability of an aether; Leonardo by his observations of waves in nature.

In Leonardo's own words, 'The May wind moves like a wave in the grain, and the wave is seen to travel over the field, but the stalks of grain do not move from their place.' This is perhaps the earliest statement by man that suggested the wave-like character of light energy and implied a stationary aether as its transmitting medium. Turning to waves on water Leonardo drew sketches of two waves moving through each other and still remaining wavelike. A straw cast on a pond would be subjected to vertical undulations caused by the wind but would remain in a fixed position. Generalizing his wave theory to include sound Leonardo stated, 'The sound from two bells remains distinct as it moves through the air, so every body gives off waves, whether light, heat, magnetism, sound, or *odor*!'

Leonardo's extension of his wave theory is interesting in that he applied the concept to invisible waves such as magnetism and heat. During Leonardo's time magnetism was a 'magical' phenomenon identified with the scarce magnetized lodestone (Fe_3O_4). Amber rubbed with silk also possessed the power to attract objects but its electrostatic 'magic' was not understood to be distinctly different from magnetism. It is also unlikely that Leonardo recognized the difference between heat transport by convection and by radiation. Yet his perception of invisible forces is clearly revealed with his observation of clouds and air, 'The air moves like a river and carries the clouds with it; just as running water carries the things that float upon it. If the clouds did not float they would be squeezed by the resistance of the air like wax pressed between two fingers.'

When Leonardo cast a pebble into a running river, he noted that the waves were elliptical, and one might claim that this was the basis for understanding the Doppler shift. This may be granting Leonardo more insight than he deserves but it is remarkable that he is credited with a clear statement of Fermat's Principle, 150 years ahead of its time, 'Every action in nature takes place in the shortest way possible.'

Leonardo challenged a common belief dating back to the Greeks that the eye was a source of light that illuminated the objects it focused on. How the Greeks could reconcile this view with the inability to see in the dark is not clear. Funnily enough, Leonardo rejected this ancient concept but for the wrong reason. He argued that there would be a time delay in observing distant objects after one opened one's eyes. He failed to appreciate that this would only be true for the eight minute time delay to the sun or the few second time delay to the moon but would not apply to the microsecond delay in the observation of a distant mountain peek.

When Leonardo sketched a camera obscura (pinhole camera) that projected an image onto translucent paper he explained it thusly, 'You will see all the objects on paper in their proper forms and colors, but much smaller. They will be upside down by reason of the intersection of the rays. These images will seem painted on the paper.' Alas! Leonardo could not accept the physiology of the eye as a type of camera obscura since he could not perceive that the images were inverted in the eyeball. His efforts at improving eyeglasses may have been hampered by this misconception. Yet he reported the contraction of the pupil under strong sunlight and he understood the stereoscopic effect achieved with two eyes. He recognized the phenomenon of persistence of vision by gazing at the stars, closing his eyes, and turning his head away.

The bending of light rays at a water-air or glass-air barrier was clearly observed by Leonardo but the concept of an index of refraction to determine the angle of bend is not mentioned in his notes. He did observe the dispersion of sunlight into its colored spectrum and correctly concluded that the blue of the sky resulted from the scattering of sunlight rather than from the intrinsic color of the sky itself.

As an experimentalist Leonardo fabricated an oil lamp that cast a brightly focused beam for working on his papers at night. The lamp employed a water-filled globe with a convex outer lens surface, while the glass holding the wick had a concave inner surface. If you sat at the correct distance from the lamp, your reading material would intersect a cylinder of light. This was one of the few ideas of Leonardo that became popular and was employed for hundreds of years.

He was probably the first to conceive the idea of a bird's-eye view of an area. By making drawings from the top of several hills, he produced such a 'three dimensional' drawing of hundreds of square miles of Tuscany.

Leonardo's notes are all written in mirror image, either to maintain secrecy, to facilitate his left hand writing, or because of dyslexia. His need for secrecy is illustrated by the incident of his assistant, 'Mirror John,' who ran off with his plans for a lens grinder and set himself up in business selling improved lenses with which to start fires from the sun's rays. At the time, lenses were fire polished after grinding, a process inferior to Leonardo's. But what the genius Leonardo knew *circa* 1500 was far from common knowledge and we can attribute an obstacle to the dissemination of his ideas to his fetish for secrecy.

One key ingredient in the advancement of optical technology during Leonardo's time was the availability of glass, the only rigid and abundant transparent material capable of bending light rays. This mating of two of the most plentiful elements in the universe, silicon and oxygen, is even today unsurpassed in its optical utility.

Its history goes back to the Phoenicians who discovered that by using native chunks of bicarbonate of soda to support their pans while heating their food on a riverbank, it reacted with the sand to form a clear fluid that solidified into glass. Glass became a popular material as both a liquid container and an art form but not until the 14th century did it become a factor in the fabrication of optical instruments.

Glass is basically a stable compound of one atom of silicon to two atoms of oxygen that can be built into a number of crystalline and non-crystalline structures. The strong electronic forces between silicon and oxygen produce its transparency since visible light is unable to break these bonds. In quartz these atoms build up a regular crystalline array while in glass they follow a more random pattern.

The addition of soluble impurities such as sodium and lead lowers the apparent melting point of glass and provides a more oven-workable material for drinking vessels and lenses. The technology associated with producing good clear glass developed in purely an empirical manner. The glass had to be cooled slowly to give a transparent product. Unless the raw ingredients were subjected to a high initial temperature the impurities that remained caused the glass to appear translucent or opaque. Once convex pieces of clear glass became available it is not surprising that serendipity led to the discovery of the magnification potential of combinations of lenses.

The glassmakers of Venice have left their mark in the beauty of their goblets and mirrors. Strict law forbade the emigration of Venetian workers in order to confine their secrets to Italy but this prohibition was impossible to enforce and the art of glassmaking spread to Germany, France, England, and Holland. Without SiO_2 it is doubtful that the age of stellar exploration would have advanced as rapidly as it did before the twentieth century although Newton did make a workable telescope with concave metal mirrors. We still have no practical replacement for silica; clear plastics lack the hardness and low cost.

A contemporary of Leonardo, Tycho Brahe, brought together glass technology and astronomical observations to invoke some order into celestial observations. This arrogant Danish scientist who was fortunate enough to secure the patronage of King Frederick II is credited with combining his telescopic observations with the construction of quadrants to accurately measure angles. In his own paraphrased words,

> We had a very large quadrant made from solid brass and very finely polished. It is five inches wide and two inches thick, and the circumference is so large that it corresponds to a radius of 194 centimeters. Every single minute of arc can be subdivided into six subdivisions; thus ten seconds of arc are plainly distinguishable and even five seconds can be read without difficulty.

The heavy device was embellished with a large portrait of himself including his disfigured nose. Tycho's arrogance had gotten him into a duel in which he lost part of his proboscis, and he replaced it with a gold-silver alloy. Nose or not, his observations in the hands of his assistant Kepler led to the determination of the earth's elliptic path around the sun.

During man's awakening to the physics of light in the sixteenth century questions began to be raised about its velocity. Such experiments as stationing two observers with shuttered lanterns on opposite hilltops, exposing one lantern, and timing the return response when the second lantern was exposed proved inconclusive. It sufficed for scientists to concede that the velocity was high.

In the century from Leonardo to Galileo the Establishment began to feel uncomfortable about the astronomical observations of the stars and planets, particularly when the heliocentric theories of Copernicus and others were put forward to displace the accepted geocentric ideas of Ptolemy. The earth was obviously the center of the universe, not the sun, and the Scriptures must be regarded as sacrosanct.

Ptolemy of Alexandria had lived a century and a half after Jesus and he recorded detailed observations of the stars in a catalog, determined the distances to the sun and moon, and even predicted eclipses. In his book Optics he pointed out that the incident ray, reflected ray, and the normal to a reflecting surface were in the same plane. He published refraction tables for light passing between air, water, and glass. The sizes of the earth, moon, and the sun became reasonably well known. Alas, Attila the Hun destroyed all that in 433 AD and the dark ages descended for half a millennium. With the rise of Christianity and the advent of the Renaissance an age of learning, including the Jesuitical orders, began to flourish. But woe to the scientist who permitted his reason to conflict with the Scriptures!

In 1572 the Church faced a dilemma. A new star appeared in the sky, surpassed in light intensity only by Venus. It shone brightly through 1573 and disappeared in 1574. This was repeated in 1604 in a different position in the sky. The plight faced by the Establishment was the clear statement in Genesis that God created the universe in six days, ending his efforts on the seventh! Such supernovae observations only added to the discomforture in Rome that the heliocentric-minded Copernicus had already produced with his 1543 publication de Revolutionibus Orbium. Copernicus escaped the fury of any central authority, although he did suffer the wrath of Martin Luther.

Where Scripture and observation appeared at variance, waffling took over. Copernicus argued that the regal nature of the stationary position of the sun and the stars was consistent with the view that immobility was nobler than movement! But when Copernicus suggested that the world was every day rushing from east to west around the sun the Establishment replied that this would obviously leave the air behind and a continual wind would be felt from west to east. Furthermore such a rapid movement would cause the earth to whirl itself to pieces, and a stone dropped from a tower would be left behind when it fell to the ground!

The infinity of the universe or, as Copernicus postulated, the immeasurable distances of the fixed stars, was taken up by Giordano Bruno who even suggested the existence of other worlds. This sort of heresy only led to his being burned at the stake in 1600. Another philosophical argument posed by a few scientists who questioned the geocentric theory, focused on the absurdity of the idea that God created the universe solely for the sake of man. Furthermore, one should not overlook that the scientific manipulation of light played but a minuscule role in the European society of the sixteenth century. Paintings, drawings, and the use of colors for clothing were far more noteworthy.

Leonardo understood color in painting when he suggested that to exactly duplicate the hues in a leaf one should dab paint on it and continue to alter the ratio of the pigments until the match was exact. He recognized the apparent increase in the size of bright objects on a dark background and instinctively understood good composition. A century ago Bernard Berenson, an outstanding art expert on the Renaissance period wrote,

> The Church from the first took account of the influence of
> color as well as music upon the emotions. From the earliest times
> it employed painting to enforce its dogmas and relate its legends,
> not merely because this was the only means of reaching people who
> could neither read nor write, but also because it instructed them in a
> way which, far from leading to critical inquiry, was peculiarly capable
> of being used as an indirect stimulus to moods or devotion.

Into this morass of observation, dogma, technology, religious teaching, and individual speculation arrived Galileo and the seventeenth century.

In 1609 Galileo's shop produced his first telescope and he embarked on his astronomical observations. While lenses for spectacles had been known since the thirteenth century and Galileo sometimes employed other people's ideas in his telescope, he always maintained that he alone had discovered the instrument. He was never a student of geometrical optics and had to rely on Kepler's understanding of refraction to construct the lens system. Galileo's practical applications of the telescope constituted his major contributions to science. Nevertheless, claiming its discovery, he sold the device to the Venetian Republic for personal gain.

He was influenced by Copernicus's writings and by his own observations to reinforce his belief in the heliocentric theory, i.e., that the earth moved around the sun. In 1610 he published *Starry Messenger*, in which he reported the discovery of the four moons of Jupiter. He named them the Medicean Planets and thus succeeded in obtaining the good will of the Medici Family.

Galileo then turned his observations to the moons of Saturn and the phases of Venus. As a result of this work, fame descended on him. This aroused jealousy within the European scientific community which found itself reluctant to accept these conclusions since the telescope was not generally available to confirm Galileo's claims. Furthermore, there was resistance within the church where it was believed that only unaided visual observations were reliable. Fortunately, Kepler dispelled some of this doubt when he confirmed most of Galileo's findings.

This did not insure that the church would accept Galileo's interpretations. The Scriptures were their last source of authority, and passages from the Bible were clearly at variance with the idea of the heliocentric theory. Nevertheless, spurred on by his sudden celebrity and always professing to be a good Catholic, Galileo attempted to curry favor with Church authorities by visiting Rome.

Though his observations were by now confirmed by Jesuit astronomers, and he was received by Pope Paul V, his interpretations led to problems. With unwise bravado, Galileo responded to the Church's claim that the moon was a perfectly smooth surface even though God made it appear as having an uneven surface like Italy.

> Really, this is a beautiful flight of the imagination. The only thing lacking in this explanation is that it is neither demonstrated nor demonstrable.

Up to this point Galileo was not considered a threat to the Church but he came into dispute with the Jesuit scientist Father Scheiner over the observation of sunspots. Galileo believed they proved that the sun rotated since the spots displayed circular motion. Scheiner claimed they were stars! (How he could believe this is difficult to fathom.)

The dispute forced the controversy between the Scriptures and Copernican theory to resurface. Clearly, the Bible stated that Joshua ordered the sun to stand still! Galileo resorted to some shilly shallying when he suggested that God used two languages, a popular one for the Scriptures and a mathematical one for science. In the former, the sun went around the earth and in the latter the earth around the sun. The theologians wouldn't accept it. In the Bible Solomon declared, 'The sun also riseth, and the sun goeth down, and hasteth to the place where he ariseth.' What could be a clearer statement that the sun moved around the earth! And so it came to pass that Galileo was called to judgement in 1616 for blowing the whistle on the Creator!

The Scriptures held the key to Galileo's error. 'The two angels had told the inhabitants of Galilee not to remain with their eyes fixed in the sky because Jesus, whose ascent they had witnessed, could no longer be seen.' The message should have been clear to Galileo and other astronomers - stop looking at the stars.

At this stage Galileo was not in deep trouble. The Church did not question his faith, but his far-reaching activities, publicized beyond Italy, became a source of embarrassment. The Holy Office officially pronounced Copernican theory heretical and Galileo was ordered to change his beliefs. Furthermore, if Galileo did not accede to these demands, he was to be imprisoned. Galileo acknowledged the Pope's order with due respect, and ended this minor ordeal with but a slap on the wrist. He returned home and prudently avoided any work that might prove to be controversial.

In 1623 Urban VIII became Pope, a man known for his scientific interests. Galileo hailed this as a good omen and believed that an era of enlightenment was about to begin. Like a scientist engrossed only in his own observations, and with all considerations of human foibles overlooked, Galileo journeyed to Rome to boast of his achievements. After six audiences with the new Pope, Galileo was blinded with optimism. Even the cautious answer of Urban VIII, when Galileo requested that the 1616 decision of the Church be revoked, failed to alert him to possible danger. Urban advised Galileo that, "If many things proved the earth moved about the sun, it was still possible that God had produced these illusions by making the sun move about the earth." Such a statement was the epitome of sitting on the fence with both ears to the ground. Galileo's self-deception was further enhanced when he heard that Urban had confided to one of his cardinals that he would not consider it heretical if Copernican theory was re-examined.

Throwing caution to the winds, Galileo returned home to initiate a campaign for reconsideration of the 1616 edict. One of the Establishment's arguments that the earth was stationary was that anything heavy fell straight towards the center of the earth when dropped. If the earth moved the object would have been given additional velocity. To further support this rationale it was contended that something dropped from the topmast of a moving ship fell ahead of the mast.

Galileo disposed of these arguments by providing a tutorial in relative motion and reporting that he had performed the second experiment with a null result.

There is an old saw that underscores the virtues of keeping one's mouth shut. Unfortunately, scientists are notorious in crowing about their findings. Galileo committed this error when he published a dialogue, a playlike discourse involving three characters, one of whom, Simplicio, extolled the virtues of the geocentric theory. The other two advocated heliocentrism and neutralism respectively. Galileo sealed his fate with this bit of playwriting, for the characters could be interpreted in many ways. Urban VIII thought that Simplicio was a veiled charade of himself. In addition political turmoil on the continent prompted the Spanish Ambassador to denounce Urban VIII for consorting with heretics. Add to this Father Scheiner's old conflict with Galileo on sunspots and you have all the elements of a powerful intrigue.

In the winter of 1632 Galileo was forthwith ordered to appear for trial. The anger of the Holy Office is seen in the summons it issued after Galileo tried to have the order rescinded or, at the very least, delayed until the spring.

This congregation of the Holy Office has taken it very ill that Galileo Galilei has not promptly obeyed the precept made to him that he come to Rome, and his disobedience is not to be excused because of the season, because it is his own fault that he is reduced to coming at this time; and he does very badly to try to soften matters by pretending to be ill, for neither His Holiness nor their Eminences are willing for a moment to listen to these fictions, or to excuse his arrival here. Therefore if he does not obey immediately, a Commissary and physicians will be sent from here to take him and to conduct him to the prisons of this supreme tribunal in chains, as up to this time he has abused the benevolence of this Congregation; and he will have to pay all the expenses thus incurred.'

And so in January 1633 at the age of 69 Galileo made the arduous journey to Rome, fully dejected and without hope. He recorded his innermost musings:

Whenever I think of it, the fruits of all my studies and labor over so many years, which had in the past brought my name to the ears of men of letters with no little fame, are now converted into grave blemishes on my reputation, giving a foothold to my enemies that they may rise against my friends and say that finally I have deserved to be ordered before the tribunal of the Holy Office, a thing that happens only because of the most serious delinquencies. This afflicts me in such a way that it makes me detest all the time I have consumed in those studies, by means of which I hoped and aspired to separate myself from the trite and popular thinking of scholars; and by making me regret that I have exposed to the world a part of my compositions, it causes me to wish to suppress and condemn to the flames those which remain in my hands, entirely satiating the hunger of my enemies, to whom my thoughts are so troublesome.

Galileo hadn't a chance. The emotional conflict totally displaced scientific arguments and Galileo was too distraught and tired to fight. He publicly declared all Copernican theory false and, on his knees with his hand on the Bible, recited his speech of atonement:

> I, Galileo, son of Vincenzio Galilei of Florence, my age being seventy years, having been called personally to judgement and kneeling before your Eminences, Most Reverend Cardinals, general Inquisitors against heretical depravity in the entire Christian dominion, and having before my eyes the sacred Gospels, which I touch with my own hands, do swear that I have always believed, do now believe, and with God's aid will believe hereafter all that is taught and preached by the Holy Catholic and Apostolic Church. But because, after I had received a precept which was lawfully given to me that I must wholly forsake the false opinion that the sun is the center of the world and moves not, and that the earth is not the center of the world and moves, and that I might not hold, defend, or teach the said false doctrine in any manner, either orally or in writing, and after I had been notified that the said teaching is contrary to the Holy Scripture, I wrote and published a book in which the said condemned doctrine was treated, and gave very effective reasons in favor of it without suggesting any solution, I am by this Holy Office judged vehemently suspect of heresy; that is, of having held and believed that the sun is the center of the world and immovable, and that the earth is not its center and moves.
>
> Therefore, wishing to remove from the minds of your Eminences and of every true Christian this vehement suspicion justly cast upon me, with sincere heart and unfeigned faith I do abjure, damn, and detest the said errors and heresies, and generally each and every other error, heresy, and sect contrary to the Holy Church; and I do swear for the future that I shall never again speak and assert such things as might bring me under similar suspicion; but I shall know any heretic or person suspected of heresy, I shall denounce him to this Holy Office or to the Inquisitor of the place where I shall find him.

How these words resonate with Nazi and McCarthy preaching!

Galileo's *Dialogue* was placed on the *Index of Banned Books*, together with the works of Copernicus, and remained there for almost 200 years. His final years were spent in blindness and he totally eschewed any controversial science. In March 1641, a year before his death, he wrote the following: 'The falsity of the Copernican system should not be doubted on any account, especially by us Catholics.'

When he died in 1642, he was refused any monument for his grave. Almost 100 years later the Church relented and permitted a tribute to Galileo provided that the epitaph did not mention Copernican theory. In 1822, the Church finally agreed to remove Galileo's *Dialogue* and Copernicus's writings from the *Index*.

We must realize that some appropriate modesty might have spared Galileo this ordeal. His astronomical observations suggest that he knew how to construct a good 30-power telescope. Scientists, such as Leonardo, were reluctant to publish such technological secrets.

Galileo's politicking at the seat of power in Rome was imprudent when it threatened the beliefs and status of men unable to grasp the significance of his findings. Scientists have not been particularly welcome in government and the Galileo story is bound to be repeated periodically, as with Oppenheimer three centuries later.

The seventeenth century advanced men's knowledge with the increased availability of books, with the development of the microscope by Robert Hooke, with the perfection by the Englishman George Ravenscroft in 1676 of a clear lead glass called flint glass, and with the publication of the monumental theories of Isaac Newton. The heliocentric theory in Protestant countries was no longer viewed with suspicion, a zeal for scientific enquiry became fashionable, and the world of art flourished in the Dutch, French, and English schools.

As early as 1629 the Rev. Francis Higginson wrote from the New World to his friends in England encouraging them to use glass for windows, a most expensive construction material and one that was taxed in England when so employed. Bottles, drinking vessels, and jewelry kept the glassmakers busy in an effort to meet public demand for this durable material.

In the area of artificial lighting the sixteenth and seventeenth centuries had advanced little since biblical days. If wood was available, particularly pitch-laden pine knots, fireplaces provided for room lighting as an adjunct to heating the room. Portable lighting relied on oils such as from olives, sheep, or fish and man had to endure the attendant odors. In the twenty-seventh chapter of Exodus we find:

'And thou shalt command the children of Israel that they bring thee pure olive oil beaten for the light to cause the lamp to burn always.'

The use of oils required the development of wicks from natural products like linen but unless the oils were clarified the capillary action could be interrupted and one had to pick hardened pieces from the wick. The lamp bodies were fabricated from clay, stone, pewter, iron and copper. It is conceivable that in the absence of adequate natural lighting at night man evolved over the millennia with the need to use the period to sleep.

Into this seventeenth century world of celestial telescopic observations, primitive lighting, and rudimentary glass technology came the genius of Isaac Newton.

Newton's contributions to mankind were partly the result of his upbringing. His father died before his birth and his mother remarried a clergyman who decided that Isaac should live with his grandmother at her farm in Woolesthorpe, England. Isaac, like many an only child isolated from other children, relied heavily on self communication. He developed a natural inquisitiveness that frequently manifested as absentmindedness. Yet he was content talking to himself, a lifelong characteristic. On one occasion he was sent to close the barn door during a storm. By the time he reached it he had become intrigued with the force of the wind and executed a series of jumps from the upper barn window, spotting his landing points. This experiment may have satisfied his scientific curiosity but it did not save the barn door from being blown off.

On another occasion he was deep in thought while leading his horse, and failed to realize that the mare had slipped its collar. He arrived at his destination devoid of horse but with his thoughts and the rein intact.

Short in stature, he nevertheless was able to soundly thrash a taller bully who goaded him into a fight, to his comrades' delight. When provoked, Isaac always gave a good account of himself.

He became intrigued with windmills and built a scaled working model of one, impressing his tutor at Grantham who recommended that he go on to university. At 19 Isaac enrolled at Trinity College in Cambridge but his limited circumstances required him to take on the job of 'Sizar', a modern errand boy catering to the more affluent students. His principal disappointment at College was the absence of natural philosophy, i.e. physics. With this subject not available, he was fortunate to find Isaac Barrow, a newly arrived mathematics tutor. He quickly took to the subject although he never would allow it to substitute for experimental observation.

As first Lucasian Professor (a time-honored seat at Cambridge), Barrow recognized Newton's innate brilliance. By the time Newton received his Bachelor of Arts degree he was well versed in mathematics. But whether it was natural philosophy or mathematics, scientific knowledge during Newton's time at Trinity was limited. It was known that glass bent light, that Galileo had produced a telescope of 30 magnification that could follow the motion of the stars and planets, and that prisms dispersed light into a spectrum of color, but little else.

In 1665 the plague descended on London and, for fear that it might spread to the provinces, Cambridge University was closed and Newton was sent back to his solitary existence in Lincolnshire. During this period he performed a key optics experiment on refranged (refracted) light through a prism. He isolated one of the colors with a slit and sent this single color through a second prism, only to find that no subsequent dispersion occurred. Newton concluded from this that light was comprised of distinct colors. He also developed the calculus (or fluxions) during this period of contemplative isolation, as well as his famous laws of motion.

On return to Cambridge, Newton was elected a Fellow of Trinity College and undertook a project that would gain him national recognition. Realizing that chromatic aberration (different colors were bent differently on passing through glass) could not be overcome with conventional glass technology, he designed a new telescope. He recognized that a metal mirror did not undergo such dispersion and cast one of speculum, an alloy of copper, tin and arsenic. He then ground and polished the surface concave and spherical (not quite the correct shape but satisfactory). The mirror was placed at one end of a one-inch diameter tube, and this focused light entering from the other end to a point halfway along the tube. A

45-degree flat mirror deflected this light out through a hole in the side of the tube. This represented a giant step forward in telescope design. With a magnification of about 40, and in the absence of chromatic aberration, he could clearly see the moons of Jupiter. A few years later a second version of this telescope was fabricated for the Royal Society and placed on public view. Charles II, founder of the Royal Society, examined it and Newton's reputation was assured.

In spite of this public recognition Newton remained a very private individual. His years alone, his inquiring mind, and his natural intelligence contributed to his unusual productivity - he did not require stimulation from scientific colleagues.

In 1669 Barrow vacated the post of Lucasian Professor and at 27 Newton was appointed to the chair. At a salary of £100 per annum and with his Fellow's stipend of an equal amount he was left comfortable. He lived and dined at College, gave a few lectures, and thought, mostly thought. What more could one wish for?

Newton was elected a Fellow of the Royal Society, an honor granted to few, and submitted a paper to them on his double prism experiment. Now in the lime-light, he found himself in dispute with Robert Hooke who claimed prior discovery of the inverse square law of gravitation. Actually this idea predated them both, but the public laundering of this debate disturbed Newton deeply and he withdrew into his shell. Hooke remained a thorn in his side for many years.

At this time someone discovered an obscure rule that required Fellows of Trinity College to take Holy Orders. Never active in theological affairs, Newton was not interested in undergoing this procedure. He was sufficiently upset over this rule to threaten resignation from the College, but his friends encouraged him to seek a waiver from Charles II. In cases like this some politicking was appropriate and Newton left for London to submit his application. Fortunately he had friends in the right places and the King, remembering the telescope, granted a dispensation.

In 1682 Newton's friend Halley discovered a comet but its apparent hyperbolic trajectory led him to conclude that it would never return. As a leading authority on planetary motion and friend to Halley, Newton suggested the alterna-tive of an elliptical, rather than hyperbolic, orbit since small segments of the two might be indistinguishable. Encouraged by this possibility Halley spent 20 years poring over old astronomical records back to 1306 and was able to confirm the comet's $75^1/2$ year period. After this study was completed the Royal Society named the celestial body Halley's Comet.

When the Catholic James II ascended to the throne in 1685 he promised to recognize the established Protestant church (the reign of Bloody Mary was still on everyone's minds) but, nonetheless, he indicated his intention to appoint a Catholic as Fellow of Trinity. A committee of Fellows, including Newton, departed for London to fight this. The King engaged the famous 'Hanging Judge' Jeffries to intimidate the committee and to defend his choice.

Meeting with Newton and his co-Fellows Jeffries bullied them into silence and contended that they were wasting their time over a minor issue. It fell to Newton to break the silence and to defy his King. Isaac made it clear that the law forbidding

Catholics at Trinity was two centuries old and was hardly a minor point since it granted the Fellow voting rights. The committee held firm behind Newton's courageous stand and the King had to withdraw the appointment. The issue received national attention and contributed to James's decision to abdicate.

Following this triumph in 1689, the University elected Newton to represent it in Parliament. However, Newton was not enough of an extrovert to be an effective speaker or lecturer.

Newton went gray before 40 and in his later years his long locks turned snow white. In addition to his virulent dispute with Hooke, he had a disagreement with Leibnitz over the discovery of the calculus. It appears they both succeeded independently.

His tome *Principia*, worked out in isolation at his farm and setting forth his calculations on planetary motion and gravitational forces, required such prodigious concentration that it left him a weakened man. It ranks with Einstein's Theory of Relativity and Maxwell's Theory of Electromagnetism as achieving the pinnacle of isolated human scientific creativity. His double prism experiment, identifying the constituents of sunlight, and his telescope free of chromatic aberration were the basis for his major contributions to optics.

Unlike Newton, his contemporary and adversary, Robert Hooke, was gregarious, was motivated through discussion with his peers, and had developed a wide circle of friends amongst all classes from chambermaids to the King. He enjoyed life's pleasures at London's coffee houses where he fraternized with his scientific colleagues. It is not surprising that he and Newton developed a hatred toward each other of such ferocity that it became a principal source of gossip amongst members of the Royal Society.

Yet some of their correspondence appeared genteel enough, as excerpted in this letter from Hooke to Newton written in 1675, a dozen years before publication of *Principia*:

> Your design and mine are, I suppose, both at the same thing, which is the discovery of truth, and I suppose we can both endure to hear objections, so as they come not in the manner of open hostility, and have minds equally inclined to yield to the plainest deductions of reason from experiment.

Newton's reply was equally non-abrasive except for a single line where he states that, If I have seen further than you, it is by standing on the shoulders of giants.' Presumably, this was a perch unscalable by Hooke!

And what was their disagreement all about? Scientific priority - a matter that still makes grown men behave like boys. Newton and Hooke both claimed discovery of the inverse square law for gravitational attraction, Hooke having published it while it still lay in Newton's notebooks for ten years. When Sir Isaac's *Principia* appeared in 1687 it failed to acknowledge Hooke; and so the years of loathing each other commenced and lasted until Hooke died. With Hooke finally laid to rest in 1703 Newton emerged from seclusion and accepted the Presidency of the Royal Society, a scientific organization that had seen much of Hooke in his capacity as Curator of experiments. In addition, Newton had withheld publication of his *Optics* until Hooke's demise and, as the new President, arranged for the Royal Society to move from Gresham College where Hooke had his residence to new quarters. The last of Hooke's ghost was thus exorcised but not his contribution to optics.

Robert Hooke was born on 18 July 1635 and spent his youth on the Isle of Wight where his brother became a grocer. His father died when he was 13 and he was sent to Westminster School in London. He rapidly assimilated Euclid and went on to Oxford where he met Robert Boyle and Christopher Wren.

Boyle engaged Hooke as an assistant after becoming impressed with his scientific comprehension. This led to his appointment as Curator of Experiments at the Royal Society, requiring him to devise three or four experiments for their weekly meetings. At the Royal Society headquarters at Gresham College Hooke encountered the notorious Sir John Cutler, a man who endowed Hooke's professorship at the College but resolutely refused to provide the money he had promised. He was continually sued by Hooke for his salary but Sir John went to his grave gloating over the fact that he never paid a farthing. In spite of that, the Cutlerian lectures became well known throughout Europe even though the benefactor contributed nought but his name.

After the Great Fire of London in 1666 Hooke's services were eagerly sought. On the recommendation of his friend Sir Christopher Wren, Hooke was appointed City Surveyor and for the next 20 years he and Wren were engaged in the replanning and rebuilding of the city. Because of the Great Fire's widespread devastation, the King welcomed schemes to revise the plan of London. Employing two points, St. Paul's and the Royal Exchange, as foci for a radial network of streets unrelated to existing roads, Wren's plan predated the ultimate layouts for Paris and Washington D.C. Hooke opted for a gridwork of avenues but neither plan was adopted.

Hooke undertook responsibility for the design of the Monument, an obelisk type structure near the site of the source of the Great Fire, with some 250 steps leading to an observation level. In addition, he planned the Royal College of Physicians and the famous hospital at Bedlam, neither of which remain standing. Not only did Bedlam provide housing for the insane but it became a tourist attraction for those wishing to view the antics of the inmates.

Of considerable impact on optics was the publication of Hook's *Micrographia*, drawings of his microscopical examinations of anything and everything. His diary recorded one of the early revelations:

> It is my wont of a morning to rub my teeth with salt, and
> then swill my mouth out with water. Yet nothwithstanding,
> my teeth are not so cleaned thereby, but what there sticketh
> or groweth between some of my front ones and my grinders,

a little white matter which is as thick as batter. I have mixed
it at divers times with spittle and then saw with great wonder
that there were very little animalcules, very prettily moving.

While the microscope has been attributed to the Dutchman Leeuwenhoek, it
was Hooke's publication with its 100 drawings that brought the instrument to the
attention of the public. *Micrographia* was avidly read by scientists and 'gentlemen
of leisure,' although its reception was not without critics. A university orator at
Oxford protested the work of the Royal Society:
'These scientists admire nothing but lice, fleas, and themselves.'
Theological debates ensued amongst the clerical members of the Royal Society. It
was suggested that the apparently functionless microbes seen under the microscope were produced by the Creator in a mood of inscrutable fun! In Hooke's
examination of fossils, he suggested that some species had become extinct but the
clerics considered it blasphemy that the Creator would make mistakes! Hooke
countered that 'the inevitable end of the world would make all species extinct.
Perhaps the Creator decided to let them die out one at a time!' The success of the
Micrographia is attested by the fact that some of the drawings were used well into
the nineteenth century.

Amongst Hooke's other accomplishments was the first observation of the
rotation of Jupiter, the hypothesis that the craters of the moon were bombardment
holes, the design of the spiral watch spring, an improved self-feeding oil lamp, and
the successful campaign to have Wren elected as President of the Royal Society
in 1681.

Hooke's diaries reveal some salacious details. Most damaging was the attention
he paid his niece Grace even while she was but a child. When she later lived with
Hooke in London as his mistress, the relationship obviously raised eyebrows. It is
not known if his brother's subsequent suicide is related to this misbehavior.
Considering the disparity in ages it is no wonder Hooke's diary constantly alluded
to Grace taking up with riff-raff.

In addition to Grace several of his housemaids served as mistresses. In one
case he discharged Doll Lord after four months with the comment 'Resolved to be
quitt of her she being intollerable!' Another mistress, Betty Orchard, suffered a
similar fate, 'She broke one of my white glasses. Threw down oyle and was
discharged after six months as being intolerably careless!'

A more lasting relationship was with Nell Young, but even here his diary
contained the entries, 'Heard of Nell lending £10 to a slut. Slept ill.' He frequently
expressed concern over Nell's all-night carouses.

This life style appeared consistent with his Bohemian existence at Gresham
College. Many items useful to a scientific bachelor were provided for him, such as
a turret for telescopic observations, apparatus and laboratory workspace, and the
Royal Society's museum and library. While he overspent for cloth for his garments,
he dressed modestly, sometimes making his own clothes. Hooke spent most of his
money on books and scientific apparatus.

In 1691 he passed the exam for MD, although he never practiced as a doctor. Curiously, in spite of his Bohemianism he never developed an interest in the theater.

By the end of the seventeenth century the world of optics was described by the *Encyclopedia Brittanica* of the period. Light was either 'an invisible fluid present at all times and in all places, but which requires to be set in motion by an ignited or otherwise properly qualified body in order to make objects visible to us.' This was contrasted to Newton's hypothesis that, 'light consists of a vast number of exceedingly small particles shaken off in all directions from a luminous body.'

The *Brittanica* argued that in employing Newton's view a luminous body would lose its particles and in time vanish! It also noted that most flowers followed the sun by some power unknown to us and this was due to the light rather than the heat. Flowers cultivated in a heated room were attracted by the sun rather than the fireplace.

Not mentioned by the *Brittanica* was the work of the Dane Ole Christensen Roemer who towards the end of the seventeenth century called attention to the ambiguity in the period of revolution of one of Jupiter's moons which changed slightly from summer to winter. This difference was attributed to the added distance the light had to travel when the earth was on opposite sides in its orbit around the sun, This took the light 18 minutes to traverse the 188 million miles of the earth's solar diameter and provided Roemer with an estimate of the velocity of light within a factor of two.

Perhaps the most important seventeenth century contribution to the advancement of optics, and to all of science, was the establishment of the learned societies. In England the Restoration brought the enlightened Charles II to the throne and the scattered groups of scientists that met informally throughout England joined together on July 15, 1662 when a royal charter was granted establishing the Royal Society. Today a magnificent painting of Charles II looks down on meetings in the main lecture hall. The exchange of ideas, the publication of papers, and the distinction of fellowship in the Society has promoted the advancement of knowledge. From its inception the nobility and businessmen were encouraged to also attend, thus lending prestige and needed financial support to the proceedings. The French and Italians soon followed suit.

The Restoration blew through England like a sea breeze and spurred intellectual thought and investigation. The Puritans had made men 'eat religion with their bread,' till the taste of it made them regurgitate. No longer did the Englishman expect to have to reconcile his discoveries with the Scriptures although the pious Newton firmly believed that science would not be at variance with ecclesiastical lore. God was taken for granted but then put aside for science and living. The causes of plagues and other natural calamities would no longer be blamed on God's retribution for man's sins. With this refreshing attitude man entered the eighteenth century. Of the five and a half million living in England in 1688 only 75,000 could be classified as engaged in the liberal arts and sciences. Their yearly income of about £60 per year was one fifth that of the 100,000 gentlemen living in England

and four times that of the million and a quarter laborers.

Yet life was regulated by the hours of sunshine. The century ended with the tallow (sheep fat) candle as the primary means to provide limited visibility at night. Even after-hours stage productions suffered from the limited light output of candles, notwithstanding the fire hazard. It took another century for man's technology to begin to make significant inroads into more efficient artificial light production.

By the start of the eighteenth century visible light sensed by the eye, and infrared heat radiation sensed by all parts of the body represented but a small portion of the electromagnetic spectrum that would later influence man in his lifestyle.

The aesthetics of light manipulation and the psychological role it played in man's psyche during the seventeenth century is illustrated by the work of Rembrandt.

Near the town of Maastricht in southern Holland there is a network of natural underground limestones tunnels. The thousands of miles of uncharted passages necessitate a guide, and during the last war the Dutch considered using the tunnels as a last holdout against the Nazis. Downed RAF pilots were smuggled into Belgium through this maze. Within this underground labyrinth were stored some of the art treasures of the National Museum of Amsterdam, to keep them out of the hands of Goering and Hitler. Most renowned of these paintings was Rembrandt's huge 'Night Watch.'

If one knew the right people during the war, it was possible to arrange a viewing. Even in times of crisis a Dutchman secured comfort by studying one of his country's treasures. This magnificent painting is the best known work of that master of light and shadow, Rembrandt van Rijn (Rembrandt of the Rhine).

The study of light can not be limited to the nature of the photon and its inter-action with matter, for Rembrandt has shown that the effective use of visible light can reveal a subject far beyond its surface. One visualizes some of Rembrandt's portraits as faces illuminated by a spotlight, but he extended this well outside the precise intensity distribution from a well-focused light source and brought an added dimension to his work. Such is the genius of an artist's imagination and technique.

The 'Night Watch' was actually entitled 'The Sortie of Captain Cocq's Company', a group of the Civil Guard that met socially, having outlived their historical function. (It parallels the Masons who no longer serve as bricklayers.) It was customary for the Civil Guard to sit for an annual group portrait, a formal row upon row of faces, all commanding equal portrait attention.

The Civil Guard of Volunteer Defense Units dated back to the time when Holland was occupied by the Spanish. In Rembrandt's time they still wore military regalia for their meetings and still played the mock soldier in their marches. The work ordered from the painter was to be large and to portray each member in full dress. The artist turned it into a theatrical display of men, costume, lighting, and movement all emerging into the bright sunshine. It depicts the confusion that reigned just before the guard was ready to assemble. At the time, Rembrandt's spouse had just died and he even slipped a child spectator into the painting, resembling closely his beloved wife Saskia.

Over the years the varnish darkened and it appeared as though the Civil Guard was making a night-time sortie, hence the misnomer 'Night Watch.' This became clear after the painting was cleaned.

There is a story that the company of men were annoyed with Rembrandt since they did not receive equal treatment as to size and prominence and they paid Rembrandt only in proportion. A close study of the work convinces one that Rembrandt's use of light and shadow was unnatural but he employed it to convey his own meaning for each member. His use of lighting is even suggestive of a halo, although this subtle point may border on irreverence.

Rembrandt's life reflected some of the tragedy that must have affected his approach to his subjects. Born in 1606 to a middle class family, his mother imbued in him a deep respect for the Bible. He was a good student and at 14 was enrolled at Leiden University. However, the lure of painting intervened and he studied the subject feverishly. During the seventeenth century artists relied on commissions, and in his Leiden years Rembrandt fared reasonably well, but the temptations of the big city attracted him to Amsterdam.

He married Saskia at 28 and she brought some wealth to the union although it contributed to Rembrandt's undoing. He relied on this money to become an avid art collector, constantly attending auctions and private sales. Though frugal in his personal needs, he became a lavish spendthrift to sustain his hobby.

Things soured considerably after he purchased a large home to raise a family and to house his acquisitions. His first three daughters died in infancy, Saskia developed poor health, and soon after, his mother died. Although he was then blessed with a son Titus, Saskia passed away shortly afterwards at the age of 30. He took a governess to look after Titus but this only ended in a breach of promise suit.

Rembrandt then brought Hendrijke Stoffels into his household and she remained his mistress until her death. It was impractical to marry her since the terms of Saskia's will decreed a cessation of her allowance if he did so. In spite of this annuity, Rembrandt found himself in financial difficulty since the payments of the house were prohibitive to a man who could not resist the auction gallery. Rembrandt had no choice but to declare bankruptcy and auction his collection at a fraction of its worth.

Hendrijke and Titus, now a man, relieved Rembrandt of his financial obligations by placing the artist in their employ and taking over his output of paintings. This legal 'bankruptcy' maneuver kept the creditors at bay, but then the church interceded by condemning Hendrijke for living in sin. She was ordered before the Church Elders and in a dramatic trial was forced to choose between Rembrandt or excommunication. She chose the latter, a weighty stigma at that time.

When Rembrandt was only 57 Hendrijke died and a few years later, Titus passed away. Tragedy forever haunted him.

During his lifetime Rembrandt executed a series of ten to twenty self portraits, all sharp contrasts of light and shade. We marvel at his effectiveness in leaving a significant portion of each work in darkness. Our brain fills in the rest, in a way that undoubtedly differs for each viewer. Our optic system is capable of storing pertinent information about thousands of faces, so that we are able to recognize people within a tenth of a second. We do not store this information as a detailed picture, since even competent portrait artists must draw faces from direct observation.

Did Rembrandt utilize some of the psychological aspects of image retention when he deliberately left parts of the subject in the darkness? One can only guess. We all fill in the shadows, perhaps only in our subconscious, but just as effectively. No one understands the genius of Rembrandt in terms of personal responses - this loosely defines aesthetics. Perhaps Rembrandt's suffering accounts for his ability to make the unseen seen. ('Supper at Emmaus' - back cover.)

Not many artists have followed Rembrandt's style, presumably for fear of being accused of lack of originality. Of all the artists over the centuries, Rembrandt, master of light and dark, has achieved the greatest popular appeal.

CHAPTER II

18th Century

While the eighteen century found western society at each other's throats with the Spanish, French, Dutch, and Americans engaged in wars over territories, men like Benjamin Franklin, Count Rumford, and William Herschel managed to play both games - scientist and soldier. During these pre-photography years the artists Gainsborough (1727-1788), Reynolds (1723-1792), and Watteau (1686-1721) popularized the fashion to secularize the medium through their portrait painting, incurring great demands on their time. The richness of colors, the novel experiment of Gainsborough's *Blue Boy*, *Pink Boy*, and the portrait of his daughters (back cover), and the growth of the Art Societies made the populace more aware of the aesthetics of form, color, and subject.

Sir Joshua Reynolds and Thomas Gainsborough became rivals in the England of the eighteenth century when their services as portrait artists were avidly sought, ensuring them financial independence. The Royal Academy of Art was founded in 1768 with Reynolds elected the first president. Through the Academy Reynolds gave popular discourses on art, in which he attempted to discover in such masters as Michaelangelo and Raphael the fundamental tenets of a grand style.

A painter should form his rules from pictures rather than from books or precepts. He who has his mind filled with ideas, and his hand made expert by practice, works with ease and readiness, whilst he who would have you believe that he is waiting for the inspirations of a genius, is in reality at a loss how to begin.

Gainsborough eschewed such high-handed postulates and established himself as a landscape painter. When Reynolds specified in one of his discourses that blue should be used sparingly, Gainsborough decided to challenge this idea and proved Reynolds wrong with his *Blue Boy*, thus raising the question as to whether art could be reduced to simple recipes. He was disappointed that his landscapes did not sell since he found difficulty in executing portraits. But by evincing none of the arrogance of Reynolds, Gainsborough succeeded both in his art and as a personality.

Dominating the 18th century as a distinguished figure was America's first physicist and outstanding personality, Benjamin Franklin, who was usually seen with his bifocals and seldom mentioned without alluding to his prowess with women. The bifocals were a clever arrangement to permit both near and far field observations. During his ambassadorship to France from 1776 to 1786 the function of the bifocals was not understood and it was popularly concluded that Poor Richard's frugality prevented him from discarding his 'cracked' spectacles. While Franklin did enjoy the company of the ladies, he was not the irresponsible roué that rumor has perpetuated. It is true that his son William was a bastard, but Franklin assumed the responsibility of raising the lad who eventually achieved status as the Royal Governor of New Jersey.

Franklin's fame in physics began with his interest in electricity. Leiden jars to store static charge were well-known curiosities, and Franklin established that electricity was both positive and negative. In his famous kite experiment (son William held the kite), he demonstrated that the spark drawn from a Leiden jar appeared identical to the electric discharge from the atmosphere. Franklin reasoned that since both discharges were similar in color, jaggedness, and noise production they must be related. The work sufficiently impressed the Royal Society that in 1756 it voted him into Fellowship, and the next year he left for London to become an active member of the Society and to represent several American states in their dealings before Parliament.

Scientifically, Franklin was also considered an authority on lightning rods and fireplace design. When St. Bride's Church was struck by lightning and lost part of its steeple, the King directed Sir John Pringle, President of the Royal Society, to suggest means to prevent recurrence. Franklin was appointed a member of the investigating committee and pointed lightning rods were recommended.

George III was so furious that Sir John consulted a rebel like Franklin that he ignored the report and forced Pringle to resign as President. He then replaced the pointed lightning rods at his Palace with the rounded ones of French design. A few weeks later lightning destroyed a wing of the Palace killing three servants. The King may have accused Franklin of 'going over his head' to make a point and the satirist Franklin probably let it be known that if he had done so, his aim would have been better. The King had never forgiven Franklin for his successful lobbying in Parliament to repeal the Stamp Tax, and the Monarch retaliated by persuading Lord North to enact a tax on tea, a dire course that inexorably led to the Boston Tea Party and the Revolution.

During his London period a clever idea of Franklin's evolved into the glass armonica. Having heard the musical glasses, which employed water at varying levels to alter pitch, Franklin conceived the idea of fabricating a series of hollow glass hemispheres, each of different diameter and pitch. These were mounted on a horizontal shaft and nested to space them as far apart as the notes on a piano. The shaft was rotated with a foot treadle and the player brought the glasses into resonance by touching the rims with moistened fingertips. This avoided the necessity of filling the glasses with water to alter pitch.

The instrument had a short-lived popularity although Mozart, Beethoven and Gluck composed for it. A quack on the continent Anton Mesmer, employed the instrument as background music to hypnotize his patients, hence the word mesmerize. Franklin was later called in to assess Mesmer's work as he denounced the man's questionable approach to medicine. (The glass armonica is currently undergoing a rebirth through G. Finkenbeiner in Waltham, Massachusetts). Franklin would often entertain his guests by singing Scottish ballads and accompanying himself on the armonica.

On his return to Philadelphia in 1775 Franklin and grandson Temple (William's bastard son whom he refused to take to New Jersey when he accepted the post of Royal Governor) discovered the Gulf Stream by taking temperature readings of the ocean. Just after signing the Declaration of Independence Franklin was sent to France to secure French aid for the Revolution. His scientific studies ceased during this period of intense political activity.

In 1786, at age 80, Franklin returned from his successful ten-year Ambassadorship in France to spend his final four years in Philadelphia. He left a unique will in 1790 that invested money for 200 years, to be awarded to Boston and Philadelphia. In 1890 a portion of the accrued investment was used to build the Franklin Institutes in Boston and Philadelphia, one of the first schools to educate technicians. Final probate of the will is expected in the 1990's having grown to several million dollars.

There is a famous bit of satire written by Franklin that reflects his sense of humor. It and other of his writings are being published by Yale University Press. It is expected that the total output of the Franklin papers will reach 50 volumes.

Advice on the Choice of a Mistress

To my dear friend:

I know of no medicine fit to diminish the violent natural inclinations you mention; and if I did, I think I should not communicate it to you. Marriage is the proper remedy. It is the most natural state of man, and therefore the state in which you are most likely to find solid happiness. Your reasons against entering into it at present appear to me to be not well founded. The circumstantial advantages you have in view by postponing it, are not only uncertain, but they are small in comparison with that of the thing itself, the being married and settled. It is the man and woman united that make the compleat human being. Separate, she wants his force of body and strength of reason; he, her softness, sensibility and acute discernment. Together they are more likely to succeed in the world. A single man has not nearly the value he would have in the state of union. He is an incomplete animal. He resembles the odd half of a pair of scissors. If you get a prudent, healthy wife, your industry in your profession, with her good economy, will be a fortune sufficient. But if you will not take this counsel and persist in thinking a commerce with the opposite sex inevitable, then I repeat my former advice, that in all your amours you should prefer old women to young ones.

You call this a paradox and demand my reasons. They are these:

1. Because they have more knowledge of the world, and their minds are better stored with observations, their conversation is more improving, and more lastingly agreeable.

2. Because when women cease to be handsome they study to be good. To maintain their influence over men, they supply the diminution of beauty by an augmentation of utility. They learn to do a thousand services small and great, and are the most tender and useful of friends when you are sick. Thus they continue amiable. And hence there is hardly such a thing to be found as an old woman who is not a good woman.

3. Because there is no hazard of children, which irregularly produced may be attended with much inconvenience.

4. Because through more experience they are more prudent and discreet in conducting an intrigue to prevent suspicion. The commerce with them is therefore safer in regard to your reputation. And with regard to theirs, if the affair should happen to be known, considerate people might be rather inclined to excuse an old woman, who would kindly take care of a young man, form his manners by her good counsels, and prevent his ruining his health and fortune among mercenary prostitutes.

5. Because in every animal that walks upright the deficiency of the fluids that fills the muscles appears first in the highest part. The face first grows lank and wrinkled; then the neck; then the breast and arms; the lower parts continuing to last as plump as ever; so that covering all above with a basket, and regarding only what is below the girdle, it is impossible of two women to tell an old one from a young one. And as in the dark all cats are grey, the pleasure of corporal enjoyment with an old woman is at least equal, and frequently superior; every knack being, by practice, capable of improvement.

6. Because the sin is less. The debauching a virgin may be her ruin, and make her life unhappy.

7. Because the compunction is less. Having made a young girl miserable may give you frequent bitter reflection; none of which can attend the making an old woman happy.

8th and lastly. They are so grateful!

Thus much for my paradox. But still I advise you to marry directly;

Your affectionate friend,
B.F.

Just as discerning is Franklin's views on the nature of light:

I must own I am much in the dark about light. I am not satisfied with the doctrine that supposes particles of matter called light, are continually driven off from the sun's surface with a swiftness so prodigious. Must not the smallest particle conceivable have, with such a motion, a force exceeding that of a 24 pounder discharged from a cannon? May not all the phenomena of light be more conveniently solved by supposing universal space filled with a subtle elastic fluid which, when at rest, is not visible, but whose vibrations affect that fine sense the eye, as those of air do the grosser organs of the ear? We do not, in the case of sound imagine that any sonorous particles are thrown off from a bell and fly in straight lines to the ear; why must we imagine that luminous particles leave the sun and proceed to the eye? It is well that we are not, as poor Galileo was, subject to the Inquisition for philosophical heresy.

This was written in 1752, well after Newton had propounded the corpuscular theory of light and, unbeknownst to Franklin, Huygens had already proposed the wave theory.

Franklin invented the bifocals during his Paris years and explains the rationale;

These I find more particularly convenient since my being in France, the glasses that serve me best at table to see what I eat are not the best to see the faces of those on the other side of the table who speak to me; and when one's ears are not well accustomed to the sounds of a new language, a sight of the movements in the features of him that speaks helps me to explain. Thus I understand French better by the help of my spectacles.

The scientific relationship between electricity and the blue light output of a spark was not queried during the eighteenth century, supposedly it was not thought that the two could be treated separately. A related phenomenon experienced by sailors in wooden ships was called St. Elmo's Fire, named after the patron saint of the seagoer. These self-sustaining balls of blue light would suddenly appear on masts and spars, dance about, and then disappear with a loud pop. No explanation exists for this well-known effect since scientists are unable to reproduce it in the laboratory. It is believed the blue color arises from ionized nitrogen.

During Franklin's period of active membership in the Royal Society a contemporary fellow of the Society John Dolland introduced an important advance in optical technology, the achromatic lens. By employing two kinds of glass, crown and flint, with differing indices of refraction a compound lens could be constructed that partially corrected for chromatic aberration, the disturbing phenomenon that led Newton to design the reflecting telescope. The Duke of Wellington, later engaged in the Spanish campaign, praised the clarity of the Dolland glass.

Of all the researchers of the eighteenth century the man who discovered much of the basic physics in optics was Sir William Herschel in his straightforward observations on the distribution of wavelengths from the sun.

He had a rather unusual start in life, his father being bandmaster in the Hanoverian Footguards. He entered the army, tasted battle at 19, found the experience not his cup of tea, and left the service. He moved to England with his brother where he continued his career as a musician, saving enough money to visit Hanover for a holiday and bringing his sister Caroline back with him. While she had hoped for a career as a singer she was smitten with the science 'bug' and became her brother's helper.

Good fortune entered Herschel's life when he discovered a new planet in 1781 and named it Georgium Sidus (George's Star after George III) with the consequent reward of a royal pension. It also brought him membership in the Royal Society and the award of the Copley Prize. Unfortunately the scientific community decided on Uranus as the name of the new celestial body, possibly because George III was beginning to show the first signs of his insanity. Nonetheless Herschel obtained royal support for a new telescope as well as a pension for his sister, the first case of British Government support for pure science. With the largest reflecting telescope in the world (40′ focal length) he experimented with various filters to observe the sun and noted that 'when I used some filters, my eye felt a sensation of heat, though I had but little light; while other filters gave me much light, with scarce any sensation of heat.'

Employing a prism to refract sunlight he positioned thermometers to receive the various colors, noting that the red end of the spectrum was distinctly hotter. While the distribution of visible light favored the yellow and green, the distribution of heat was clearly maximized in the *invisible end beyond the red*. Herschel tested these heat rays and found they were reflected from a mirror as was the visible light but he still concluded that heat and light were different 'animals'.

During this period of time, when Herschel and Franklin enjoyed membership in the Royal Society, an American-born scalawag Benjamin Thompson entered the scene, and added 'color' as well as knowledge to the subject of heat radiation.

In the early nineteenth century Count Rumford (Benjamin Thompson) was invited by President John Adams to return from London and become the first Superintendent of the newly created US Military Academy at West Point. Rumford welcomed this distinguished honor but shortly afterwards withdrew his acceptance. During this brief interval it had been brought to President Adam's attention that this scientist, famed for his work on heat and radiation, had been a spy and traitor to the American Revolutionary cause. In fact, this was only one of an innumerable number of incidents that revealed that Rumford was virtually amoral. He left his mark of villainy in America, England, France and Germany. Of all the scientists with international repute Rumford proved to be the greatest rogue of all time.

Sympathy may be elicited for Benjamin who was born into a family with some expectations. His grandfather was a gentleman of means but Benjamin's father died before his grandfather, and the right of inheritance passed to the younger brother, Benjamin's uncle. As a young teenager Benjamin's life was suddenly turned about, particularly when a cruel stepfather joined the family and ignored Benjamin's eagerness to attend school. To avoid continual wrangling between the two, Benjamin's mother arranged for his apprenticeship to a dry goods dealer in Boston.

The lad's natural curiosity found him toying with chemicals and rocketry. On being left in charge of the shop one afternoon, Benjamin set off a rocket that landed on the roof and burned part of the store to the ground. This ended the apprenticeship and sent Benjamin home to a good tanning by his stepfather, comforted only by the chemistry book hidden in his breeches. A second apprenticeship was secured with a doctor who recognized Benjamin's natural intelligence. By age 17 the tall, handsome, red-headed lad had absorbed everything in the doctor's library, learned French and the violin, and discovered that one could kill a pig (but not roast it) by tying a kite to its tail in a thunderstorm.

At 19 Benjamin completed his apprenticeship and took his violin and meager belongings to Rumford, New Hampshire, to become a school teacher. The recently widowed Reverend's daughter, who had inherited a magnificent estate, brought the lad under her wing and, though 11 years his senior, was smitten and proposed to him. Of course the tongues clacked, but the ensuing marriage transformed this poor lad into the local squire. Benjamin decided that this warranted an appropriate pedigree and he began boasting descent from such notables as the Earl of Suffolk in the Court of Charles I, and even designed an impressive coat of arms to support his claim.

Young Squire Thompson began experimenting in agriculture and animal husbandry and his efforts caught the attention of the Royal Governor John Wentworth who engaged Benjamin in a survey of New Hampshire. Benjamin performed his duties in a magnificent coach proudly bearing his 'new' coat of arms.

With the mounting concern over differences with London, Benjamin was appointed a Major in the New Hampshire militia, the youngest to hold that position. The British had occupied Boston in 1774 as punishment for the Tea Party but a number of the troops were deserting, lured by the prospects of cheap land. Major Thompson was instructed by the Royal Governor to do some spying and report on these desertions. After a number of soldiers were caught and hanged, Thompson was accused of treachery by the Americans and brought to trial.

Leaving his wife and child behind with the parting shot that, 'She married him, not the reverse,' Thompson escaped and reached British Headquarters in Boston. General Gage asked Thompson to infiltrate Washington's Headquarters just outside Boston and report on troop disposition. Using secret inks, he provided Gage with this intelligence, but was eventually caught and brought to trial. He again escaped to the British lines and left for England when Washington forced the

British evacuation of Boston on 17 March 1776.

Armed with a letter of introduction from General Gage to Lord Germain in London, Thompson paraded himself off as a full colonel and joined Germain's staff, charged with provisioning the troops in America. As part of his duties, Colonel Thompson turned to experiments testing the efficiency of various gunpowders by employing a ballistic pendulum, a barrel of sand freely suspended from the ceiling that recoiled in proportion to the momentum of the bullet fired into it.

At the time, Lord Germain was in the midst of a political dispute with Lord Sandwich, the First Sea Lord, and he prevailed on Thompson to do some spying, a task Benjamin felt uncomfortable with. Under pretext of examining the range of naval guns, Thompson was invited aboard Lord Sandwich's flagship. Taking copious notes on the inefficiency of the seamen and details of poor naval operations, Thompson provided Germain with a potentially unfavorable report. Such was the unchanging political backstabbing of the time!

After reporting on his experiments with the ballistic pendulum, Thompson was elected to Fellowship in the Royal Society at the early age of 27. With this added honor Colonel Thompson developed a personality of unparalleled arrogance, particularly when Germain placed him in charge of outfitting the troops in North America. Benjamin performed some experiments on thermal conductivity, convincing himself that wool was best for keeping soldiers warm in winter. (Unfortunately it was also the worst choice during the sweltering summers in America.) With a job paying so handsomely it is inexplicable to account for Thompson's next move, the decision to consort with the French.

In several secret meetings in London he offered to sell military information. The French spy was caught and Lord Sandwich accused Germain of harboring a traitor in the person of Thompson. Benjamin denied this and was fortunate that the Frenchman, even under extreme torture, did not reveal Thompson's identity. Sandwich nonetheless threatened to expose the young Colonel, whereupon Germain hinted that he would reciprocate by publishing Thompson's report on the Fleet's inefficiency. The stalemate was resolved when Thompson volunteered to form a company of British militia and return to America to engage in the conflict.

Colonel Thompson's regiment undertook maneuvers around Huntington, Long Island, where he ruthlessly destroyed property. He confiscated tombstones and used them to bake bread, selling loaves to the residents with R.I.P. baked in reverse in the crust. Fortunately for Long Islanders the war ended before Thompson could vent his spleen any further. (For some bizarre reason Huntington has a Count Rumford society.)

Returning to London, Germain decided Thompson had become an embarrassment and recommended his services to the King of Bavaria who was looking for someone to whip his small army into shape. Thompson agreed to spy on the Bavarian army for the English if Germain obtained a knighthood for him. George III was not pleased with this *quid pro quo* but went along with Germain's recommendation, hoping he would see the last of the arrogant scientist.

Meeting King Theodor in Munich, Thompson boasted of his military prowess and was appointed General with the task of bringing the army up to standard. Thompson undertook some reforms that made him a hero. He was shocked by the large beggar community in Munich and, with the King's approval, rounded them up and placed them in an institution in which they were set to work making uniforms for the soldiers. In order to feed this group Thompson turned the soldiers to farming. In a scientific study Thompson discovered that a soup made of potatoes and pulses, flavored with onions and vinegar, was the cheapest and most satisfying food for the beggars. Even today, Rumfordsuppe is known to Bavarians, although not consumed.

Following his success in organized socialism, Thompson engaged in his famous gun-boring experiments from which he argued against the caloric theory of heat. After a few years in Munich, he decided to return to London and report his findings to the Royal Society. However, Germain was so irked that Thompson had never provided any spy reports on the Bavarian Army that he arranged to have his carriage waylaid as it passed through St. Paul's Churchyard, relieving Thompson of all his research papers. It was Germain's unfulfilled hope that these would contain military intelligence.

Thompson presented his experimental work, as best he could, to the Royal Society and became embroiled in controversy. It was commonly believed that bodies raised their temperature by absorbing a substance called caloric. Thompson argued that heat was an internal state of matter and cited his gun boring experiments under water. If the gun barrel absorbed caloric from the water, then the barrel would cease being heated after the water had lost all its caloric. However, claimed Thompson, his experiments showed that the barrel would continue to heat as long as the boring operation continued. Since his experimental notes had been stolen from him, the debate went unresolved.

In the early part of the nineteenth century wood was scarce in both Europe and America prompting Rumford to redesign fireplaces to increase their heating efficiency. By reducing the size of the throat and elevating it significantly the efficiency was easily doubled. Funnily enough, very few modern fireplaces have adopted the Rumford design although during his lifetime his advice in this matter was eagerly sought. (Recent measurements show the Rumford design to be at least twice as efficient as conventional fireplaces.)

Thompson returned to Bavaria and decided to help the residents of Munich by providing a large open space for walking, gardening and recreation. Such was the King's respect for Thompson's ideas that he gave up his private hunting preserve, the only undeveloped land in the city. With the planting of formal gardens the park became known as the English Gardens and so remains today, one of the largest urban parks in Europe.

Thompson's ego was continually fed with successes, and his arrogance waxed. The King's Councillors grew jealous of this foreigner and raised objections to Thompson's plan to circulate a petition amongst the citizens thanking the King for the Gardens. Not only did Thompson convince the King that his Councillors were behaving disrespectfully but even suggested a public punishment. The Councillors, one by one, apologized to the King by kneeling before a *painting* of Theodor as the entire populace watched and jeered. That was the last straw and the Councillors threatened mass resignation. The dispute almost came to an abrupt halt when Thompson accidentally blew up his laboratory during an experiment on heats of combustion. The King sent Thompson to Italy to recuperate and, to show his gratitude, provided Thompson with a pair of his mistresses for comfort. They were, in fact, sisters.

Upon his return to Munich the King invested Thompson as Count of the Holy Roman Empire, a title reserved for nobility. That was the good news, the bad news was the King's decision to have Thompson leave Munich. It was either Benjamin or his Councillors! To dignify the banishment the King appointed Count Rumford Bavarian Ambassador to the Court of St. James. The name Rumford had been selected by Thompson from the town in New Hampshire in which Benjamin had spent his youth.

Knowing how King George III felt about Rumford, King Theodor deliberately ignored protocol by failing to notify the British Court in advance of the identity of the new Ambassador. On Rumford's arrival in England the King refused to recognize the appointment and Rumford found himself up the creek without a paddle.

He was rescued from this situation by an urgent call from King Theodor to return to Munich. The French and Austrian armies, both far outnumbering the Bavarians, were prepared to meet in conflict and the battleground was likely to be Munich! The King's Councillors were certain that Rumford would be left in ignominy by the two large armies and this would destroy his prestige, hence they were eager for his return to Munich as the army General. Rushing back to Bavaria, Rumford arrived to find the King, his Councillors and much of the populace evacuating the city. Count Rumford and the army were left in charge.

After organizing the defenses, Rumford prepared for the arrival of the two armies. The French and Austrians camped outside the city walls and, while they awaited battle the next day, Count Rumford approached the Austrian Commander under a flag of truce and convinced him it was not necessary to engage the French within the city walls. He then met with the French General, who was impressed with Rumford's honors from the French Academy of Science, and he gained the same concession. Munich was saved!

When King Theodor and his Court returned, Rumford's success only infuriated the Councillors further, and the King suggested a long holiday for Rumford - until tempers cooled. It was at this point that Rumford wrote President Adams to suggest starting a Military Academy at West Point. Rumford left for England hoping the proposal would be accepted.

On arrival in London Rumford was visited by the American Ambassador and presented with the letter of appointment as first Superintendent of West Point, only to be asked that after a respectable delay he should decline the post in order to avoid the embarrassment that would accompany exposure of his previous misdeeds as a spy.

Rumford turned his attention elsewhere and conceived the idea of an institution to display the latest achievements in science. He received approval from the Prince Regent, later George IV, and selected a spot on Albermarle Street. Unfortunately the owner of the land would not sell. How lucky for Rumford that the owner was killed by a highwayman a week later! (Perhaps you're thinking the same thing we are.) The bereaved widow quickly accepted Rumford's offer to purchase the land.

When the Royal Institution was completed a few years later under his direction, Rumford only permitted his own work on heat and radiation to be displayed. He judged no other scientist as worthy! He also made a sizeable donation to the Royal Society to pay for a gold medal to be awarded annually for achievement in science. After accepting the gift, the Royal Society made Rumford the first recipient, mostly to keep him out of their hair in future deliberations on awards. The medal is still given annually in England and in America, the latter through a Rumford donation to the American Academy of Arts and Science.

To everyone's delight at the Royal Institution, Rumford announced that he was returning to Munich to build a similar institution. On the invitation of Napoleon he stopped in Paris en route. Here he met Madame Lavoisier, whose famous husband had been guillotined during the Revolution. She was the leading lady of Paris and Rumford invited her to join him on his trip to Munich. In Bavaria she became so impressed with his achievements that she proposed marriage. They returned to Paris for a long honeymoon but the two strong personalities soon clashed. Actually, were it not for his marriage, Rumford's previous duplicity in spying would have made him *personna non grata* in France. After several good squabbles Rumford parted company with Madame Lavoisier and spent his few remaining years in Paris.

Under the terms of his will Rumford endowed a Chair of Physics at Harvard in return for Harvard's tending his grave at Auteuil. The present bearer of the Chair confessed to me he knew nothing about Rumford. The town of Rumford later changed its name to Concord in order to disassociate itself from the traitor. Thus ended the life of a man whose arrogance assured him a place in oblivion.

Rumford's expertise in heat and radiation placed him in great demand for designing fireplaces. He understood the fundamentals of heat and its movement by conduction, convection, and radiation, although he probably did not understand that radiation was infrared light.

Today the Royal Institution is a respected name in London and it has done much to popularize science. Its lecture theatre has been altered but slightly since Rumford's time, boasting excellent acoustics and sightlines. The Englischer Garten in München provides recreation for tens of thousands of Müncheners and the Rumford Medal is still prized as is the Rumford Chair.

The unfortunate turn of events in his youth may have contributed to his unbelievable display of arrogance. Other than Rumfordstrasse in Munich and a plaque in the Englischer Garten, the Bavarians have no idea of this man's contribution to their city. The poor boy from Woburn, Massachusetts, achieved the impossible by becoming a Count of the Holy Roman Empire but his arrogance has relegated him to a position of obscurity.

One other development in optics during the eighteenth century was the appearance of street lighting in the large cities. The demand for whale oil, the most efficient lighting fuel, sent whalemen to the far corners of the world hunting the mammals. While whale oil exceeded tallow for light output the danger of oil spills and fire had to be reckoned with. The Great Fire of London in the seventeenth century was a lesson no one would forget!

When whale oil was unavailable or too expensive the housewife would resort to technologies passed on from previous generations. These included the use of wax candles from honeycombs, tallow from animal fat, and lovely scented bayberry wax. Candle dipping was a tedious process that was often relegated to the children.

Candle holders were fabricated from a wide variety of materials including mirrors to reflect light from a wall sconce.

A new and sometimes masterly and clever use of optics emerged during the eighteenth century with the work of political cartoonists led by James Gillray. Within a single line drawing and the 'bubble' technique for characters mouthing their lines, the cartoonist attacked the follies of those in power, surprisingly immune from any charge of sedition or defamation. Even the King became a target for ridicule.

The best known cartoon of this period was executed by Paul Revere following the Boston Massacre of 1700 (back cover). It was effective in arousing Colonial antagonism toward the British and the forced quartering of soldiers. It is a gross distortion of the facts, the redcoats having never been ordered by Captain Preston to fire at the citizenry, the soldiers having been provoked by rock throwing, and the soldiers not massed in line formation. When this cartoon was circulated amongst the colonies the citizens of Boston educed the sympathy of the colonials. John Adams, a Braintree lawyer, successfully defended Captain Preston and the soldiers in the subsequent trials. No matter, the event had aroused the fury of the Bostonians and forced the British to remove the soldiers until the Tea Party three years later.

A well-known anti-recruiting poster of October 1775 exposed the menial pay of the redcoat. The emaciated recruit is implored by his naked children and pregnant wife not to go to war. His pay of a farthing a day (one 48th of a shilling) is compared to that of a coachman at three shillings, a driver at two shillings, and a chimney sweep's boy at one shilling. The recruit is beckoned by the gaunt figure of Famine. Many of the redcoats deserted on arrival in America when they discovered that land was cheep, every farmer being self-sufficient.

An amusing Gillray cartoon of 1793 depicts the threat of a French invasion being repulsed by George III with a blast of excrement. Another Gillray cartoon of 1793 parodies the French Reign of Terror and the guillotining of Louis XVI. Cartoonists effective employment in coupling a visual pattern with the individual's emotion is still one of the most efficient of such techniques.

To sum up, the eighteenth century ended with a modicum of understanding about the basic physics of light, mostly through the work of Herschel and the previous efforts of Newton and Hooke. Very few engaged in its study. Looking back it was a painfully slow process in creating those conditions of peace and mass education that aroused common interest in the subject. The fiercest debate to engulf the subsequent century was between those espousing the corpuscular theory and those favoring the wave theory. Of course, both were right but it took still another century to realize that. Even in my own undergraduate days before World War II the dichotomy was still bandied about. Man cannot easily envision the submicroscopic world that he cannot see, hear, nor touch - a perpetual barrier to our understanding.

CHAPTER III

19th Century

The nineteenth century found those countries eminent in scientific research embroiled in belligerent activity. England and the German States were fending off the brilliant Bonaparte as he marched through Europe. Yet in spite of the French Revolution and the uncertainty of France's future during Napoleon's rapid rise to power, his election to the French Academy made him a lifelong friend of science, although not the arts. Whenever he could, he attended their meetings, lending moral support to the subject. He announced a grand prize for anyone able to achieve discoveries comparable to those of Volta and Franklin. He wooed Rumford when the famous rogue visited Paris and courted Madame Lavoisier. However, while Frenchmen like Laplace, Dulong, and Poisson came to the fore during Napoleon's reign it was the Englishman Thomas Young who provided the next giant leap in optics research at a time when the wave and corpuscular theories were engaged in their own war and the physiology of the eye began to attract keen interest amongst physicists.

Thomas Young was unquestionably a genius. Born into a Quaker family in 1773 in the town of Milverton, England, he began reading at the age of two and finished the bible twice by the age of four. He later achieved fluency in 14 languages and excelled in several disciplines.

The Quaker work ethic sent him on a course of diligent self-study with but minor interruptions for leisure activities. In Young's day dancing, art, and even literature were banned from the sect and the Quakers became known as the 'truth seekers' - they contributed their fair share to the rolls of the Royal Society.

At 13 a relative took him to London and young Thomas avidly browsed through the book shops. Asked to replace an expensive tome that one bookseller considered too difficult for such a youth, Thomas was not intimidated. The bookseller smiled and told Thomas he might keep the book if he translated some of its arcane language. The youth walked away with his prize.

Recognizing his genius, David Barclay the banker invited Thomas to become companion to his grandson and the two lads became absorbed in a study of languages and in the works of Newton, Linnaeus, and others on chemistry, astronomy, mathematics, philosophy and botany. Thomas's study of calligraphy developed into a handsome handwriting. Over the five-year period of friendship with the Barclay grandson, Thomas wintered in London, enjoying the comforts of the famous Quaker family.

Under the influence of his uncle, Richard Brocklesby, Thomas opted for training in medicine and undertook studies in London and Edinburgh, entering St. Bartholomew's Hospital in 1793 where be became absorbed in the physiology and focusing property of the eye. He was sufficiently productive in his scientific output to be elected a Fellow of the Royal Society at 21.

In Edinburgh he learned practical anatomy through dissection of corpses, although he found himself in the midst of a scandal over body snatching. The school's records on the matter are preserved in the censure meted out by the Principal:

> Of late there has been a violation in the Greyfriars' churchyard by some who most unchristianly have been stealing, or at least attempting to carry away, the bodies of the dead out of their graves. But that which affects the school the most is a scandalous report, most maliciously spread around the town, that some of the doctors are accessory, considering that the magistrates have been willing to allow the school those dead bodies that come under their jurisdiction.

This latter point referred to those who had been hanged and were handed over to the school. As a consequence of this violation of the sanctity of the grave, the Anatomy Department was forced to move out of town.

Emulating Benjamin Franklin, Young set a rigid schedule of training for himself to cover his intended accomplishments at various ages:

2 Reading English
6 Latin and writing
8 Arithmetic
10 Greek
12 French
14 Italian, geometry
16 German and mathematics
17 Natural philosophy, drawing
18 Chemistry, biology

This schedule actually fell short of the many exotic languages he learned while absorbing the subjects necessary to pass his medical examinations.

After completing his studies in Edinburgh he toured Scotland on horseback becoming enamored of the high life in the city and deciding to abandon the Quaker custom of eschewing the arts by becoming a Sarah Siddons *devotee*.

In 1795 he travelled to Göttingen, and attended the new university founded by George II, passing the examination as an MD. He was prevented from embarking on the Grand Tour around Italy and to Vienna due to the war between Britain and

Austria, and he limited his holidays to a tour of the Harz Mountains with Thomas Wedgwood, son of Josiah.

On his return to England, he entered Cambridge (Emmanuel College) and, in order to fulfill the requirements for an English MD, totally abandoned his Quaker faith by taking an oath to the Church of England. As a member of college he became President of the Parlour Club, a group dedicated to intellectual and other forms of entertainment with frequent bets to enliven the proceedings. Young's minor idiosyncrasies began to emerge, notably a frankness that could offend. As a scientist and Doctor he insisted that his ideas received due credit.

He was loath to accept some of the unreliable medical practices of the day, often quoting a poem about a Quaker doctor, I. Lettsom:

'When any sick to me apply
I physics, bleeds and sweats 'em;
If, after that, they choose to die,
Why verily,
I. Lettsom'

With the demise of his uncle Brocklesby, the inheritance of his London house in Park Lane and £10,000 cash, Young became financially independent, although hardly rich. His reputation at the turn of the century, when he accepted a professorship at the Royal Institution and published his Royal Society paper on light and sound, began to achieve international proportions. He continued to practice medicine without bleeding patients, claiming that more people were killed in the eighteenth century by the lance than the sword.

He bought a place in Worthington on the south coast where bathing became fashionable, and he completed the picture of a respectable member of the community by marrying Eliza Maxwell in 1804, a woman about whom very little was known at the time or has subsequently come to light. He continued his medical and scientific interests but never achieved success as a doctor, due mainly to his candor and lack of tact with patients, although his percentage of cures was higher than for most doctors. He lectured (not too well) at the Middlesex Hospital, was elected to a Fellowship of the Royal College of Surgeons and joined the staff of St. George's Hospital.

At the end of the nineteenth century his occupation with scientific interests produced the eriometer, a device for measuring the size of small particles by relying on the diffraction of light through tiny holes and comparing this to the particle of interest. The device was not fully understood by Young at the time although it did yield accurate values for the size of red blood cells, fine seeds and the thickness of wool fibers. These all have the dimensions of microns (ten thousandths of a centimeter), the size of near-infrared photons.

Between 1800 and his death in 1829, Young delved into a broad variety of problems and is considered the father of physiological optics. The question of how the eye adjusted itself to view objects at varying distances aroused his attention. He recognized that the eye was capable of adjustment from a 'far point', which could be infinity for those with normal vision, to a 'near point' around ten inches. Below this distance the eye struggled to bring objects into focus. To establish the near and far points he invented the optometer, a simple device that consisted of a card with two holes through which a black dot was viewed. If the dot was at a distance such that it appeared single, then the eye had been able to accommodate. At shorter or longer distances, though, the image on the retina appeared double, placing it outside the accommodation range.

In the course of determining the near and far points, Young discovered that they would vary depending on whether the holes in the card were held horizontal or vertical, hence he coined the term astigmatism. During these studies he identified chromatic aberration, an astigmatism due to differences in the index of refraction for different colors, and spherical aberration, an astigmatism that varied circumferentially around the eye.

In a series of experiments on himself employing lenses and underwater observations, Young concluded that accommodation did not result from changes in the curvature of the cornea or changes in the length of the eyeball. Rather it was due to changes in the crystalline lens shape, an elastic, jelly-like body behind the pupil.

He went on to consider color perception. In his own words:

> Now, as it is almost impossible to conceive each sensitive point of the retina to contain an infinite number of particles, each capable of vibrating in perfect unison with every possible undulation, it becomes necessary to suppose the number limited, for instance, to the three principal colours, red, yellow, and blue, of which the undulations are related in magnitude nearly as the numbers 8, 7 and 6; and that each of the particles is capable of being put in motion less or more forcibly by undulations differing less or more from a perfect unison; for instance, the undulations of green light being nearly in the ration $6\,^1/_2$, will affect equally the particles in unison with yellow and blue, and produce the same effect as a light composed of these two species; and each sensitive filament of the nerve may consist of three portions, one for each principal colour.

In subsequent lectures before the Royal Society he altered the three principal colors to green, violet and red. Young recognized that color perception resided in the physiology of the eye rather than in the nature of light, i.e. the eye's receptors had fluorescent organic substances that responded to a narrow range of colors. In color-blind individuals one or more of these receptors is inoperative.

Young's career at the Royal Institute began when he received an offer of a professorship from the founder Count Rumford at a 25 per cent reduction in salary from that of his predecessor, and with an increased workload. Young's reply is preserved:

> I confess I think it would be in some measure degrading both to me and the Institution that the salary which appears to me to

have been no more than moderate before, should now be reduced by one fourth, at the same time that the labour and responsibility of the employment are rather increased than decreased.

The very next day the managers of the Institution recommended a higher salary, Dr. Young's reputation considered commensurate with the undertaking.

Unlike Davy, Young's lectures were too highbrowed for the broad spectrum of attendees at the Royal Institution's Friday evening talks. In fact, the popularity of Davy's lectures had already produced such traffic jams that Albermarle Street was forced to become the first one-way street in London.

Young turned his attention to a comparison of the dynamical theory of heat with the wave theory of light. In the early nineteenth century phlogiston (or caloric) was considered to be a separate substance responsible for heat. Both Young and Rumford recognized that the origin of heat resided in the internal dynamics of matter. In Young's own distinctive verbosity:

If heat is not a substance, it must be a quality; and this quality can only be motion. It was Newton's opinion that heat consists in a minute vibratory motion of the particles of bodies and that this motion is communicated through an apparent vacuum by the undulations of an elastic medium which is also concerned in the phenomena of light. If the arguments which have been advanced in favour of the undulatory theory of light be deemed valid, there would be still stronger reasons for admitting this doctrine respecting heat; and it will only be necessary to suppose the vibrations and undulations principally constituting it to be larger and stronger than those of light, while at the same time the smaller vibrations of light, and even the blackening rays derived from still more minute vibrations may, perhaps, when sufficiently condensed, concur in producing the effect of heat. These effects beginning from the blackening rays, which are invisible, are a little more perceptible in the violet, which still possesses but a faint power of illumination; the yellow-green afford the most light; the red gives less light, but more heat; while the still larger but less frequent vibrations, which have no effect upon the sense of sight can be supposed to give rise to the least refrangible rays and to constitute invisible heat.

Young's greatest contribution to optics came with his explanation of the diffraction of light by long narrow obstacles. Young calculated the fringe spacing in the diffraction pattern from such obstacles. But after publishing this paper he was chagrined to read the following editorial tirade in the *Edinburgh Review*:

We now dismiss, for the present, the feeble lucubrations of this author, in which we have searched without success for some traces of learning, acuteness, and ingenuity, that might compensate his evident deficiency in the powers of solid thinking, calm and patient investigation, and successful development of the laws of Nature, by steady and modest observation of her operations. We came to this examination with no other prejudices than the very allowable prepossession against vague hypothesis, by which all true lovers of science have for above a century and a half been swayed. We pursued it, both on the present and on a former occasion, without any feelings except those of regret at the abuse of that time and opportunity which no greater share of talents than Dr. Young's are sufficient to render fruitful by mere diligence and moderation. From us, however, he cannot claim any portion of respect until he shall alter his mode of proceeding or change the subject of his lucubrations; and we feel ourselves more particularly called upon to express our disapprobation, because, as distinction has unwarily been bestowed on his labours by the most illustrious of scientific bodies, it is the more necessary that a free protest should be recorded before the more humble tribunals of literature.

That the editor of the *Edinburgh Review* possessed no qualifications as a physicist did not expiate the damage to Young's reputation. After further similar vituperation Young published a verbose reply that sold but one copy; it was not the practice at the time for journals to accept letters of rebuttal to the editor.

A man who has a proper regard for the dignity of his own character, although his sensibility may sometimes be awakened by the unjust attacks of interested malevolence, will esteem it in general more advisable to bear in silence, the temporary effects of a short-lived injury, than to suffer his own pursuits to be interrupted, in making an effort to repel the invective, and to punish the aggressor. But it is possible that art and malice may be so insidiously combined as to give to the grossest misrepresentations the semblance of justice and candour; and, especially where the subject of the discussion is of a nature little adapted to the comprehension of the generality of readers, even a man's friends may be so far misled by a garbled extract from his own works, and by the specious mixture of partial truth with essential falsehood, that may not only be unable to defend him from the unfavourable opinion of others, but may themselves be supposed to suspect, in spite of their partiality, that he has been hasty and inconsiderate at least, if not radically weak and mistaken.

Young's successful demonstration of the diffraction by a narrow obstacle yielded good values for the wavelength of red and violet light and left his scientific critics behind. Yet one item that plagued him was the double refraction of light in crystals of Iceland spar, a substance that breaks light into two rays because of its non-cubic crystal structure. The clue to this behavior depended on the recognition that light could be polarized, a phenomenon requiring a transverse wave theory, i.e. light vibrates at right angles to its direction of propagation, the index of refraction depending on the angle of the polarization direction relative to the crystal direction.

In the years until his death in 1829 his interests were occupied by a number of topics. As a linguist he was the first to crack the code on the Rosetta Stone, discovered in 1799 by French troops excavating some ground in the Nile delta. The famous stone contained a running text in three scripts, Greek, demotic and hiero-glyphic Egyptian. Being damaged, though, it could not be literally translated and Young performed some word analysis to identify the word '*Ptolemy*' and start the scholars on the road to its decipherment.

Young was called upon to contribute to the *Encyclopedia Brittanica* and he provided articles on everything from bathing to double refraction and tides, the latter involving original calculations.

A medallion to Young's memory has been placed in Westminster Abbey:

SACRED TO THE MEMORY OF
THOMAS YOUNG, M.D.,
FELLOW AND FOREIGN SECRETARY OF THE ROYAL SOCIETY,
MEMBER OF THE NATIONAL INSTITUTE OF FRANCE:
FIRST ESTABLISHED THE UNDULATORY THEORY OF LIGHT
AND FIRST PENETRATED THE OBSCURITY
WHICH HAD VEILED FOR AGES
THE HIEROGLYPHICS OF EGYPT.

His name is permanently associated with the Young's modulus, the fractional amount by which a material is distended under a given stress.

Young's ability to explain diffraction by the addition of waves led him to abandon Newton's corpuscular theory but Young's proposal that light was longitudinal in character, i.e. the wavelike vibrations were in the same direction as its direction of propagation, hindered him in explaining double diffraction in Iceland spar. Young and Laplace fought over this phenomenon, Laplace claiming that within the framework of the corpuscular theory light was split into two components of different velocities.

When Etienne-Louis Malus in France discovered that the two different rays in Iceland spar did not interfere with each other he suggested the name polarization for the phenomenon and won a prize for his work. Malus is another example of a scientist who became involved in military matters, serving under Napoleon in the Egyptian campaign. At Jaffa the French were overwhelmed by the enemy and all officers were beheaded except Malus who lay undetected asleep in a trench. His major discovery occurred in Paris when he used Iceland spar to examine the sun's rays reflected from a window in the Palais du Luxembourg. He determined that the two states of polarization (they were named the ordinary and extraordinary rays) were not of equal intensity. That evening he found he could completely polarize light from a candle by reflection from a mirror at 36 degrees.

The polarization of light did not explain diffraction and at this stage Fresnel came forth to resurrect the wave theory in a slightly differing form than Young's. Afterwards the secretary of the French Academy encouraged Fresnel and this led to Fresnel's proposal that light was transverse - it vibrated at right angles to its direction of propagation, capable of being split into two non-interfering polarization states ninety degrees apart.

Fresnel made an interesting contribution to optical technology through his construction of French lighthouses in the 1820's. He designed a lens with annular segments of different focal length leading the secretary of the French Academy to declare that such a design offered as much illumination as one third of all the street lamps of Paris combined and provided France with the most beautiful lighthouses in the world.

Contemporary with the work of Fresnel in the 1820's was the innovation of the Scotsman David Brewster who created the kaleidoscope, an instrument that was solely employed to entertain. In its first months some 200,000 were sold in Paris and London and America. Alas, not having patented the instrument Brewster reaped nothing from the invention, particularly exasperating for a Scotsman.

While it was inconceivable at the time to expect scientists to visualize light as excessively complex - although the wave and corpuscular pictures could be easily imagined within the framework of man's experience but not incorporated into a single theory. As we know, light lives in its own world which combines the two concepts into a composite picture for photons. The world of the photon, the electron, and the atom has its own rules of behavior. The attempt by man to produce illustrative models that behave like things do in their macroscopic world may provide one with a sense of familiarity but it can easily mislead. The history of science reveals that new discoveries always raise more problems than they solve.

The next nineteenth century mystery to bend to a plausible explanation was solved by Fraunhofer when he examined the spectrum from the sun with a more accurate dispersion instrument. He accomplished this by first passing the sunlight through a slit before it reached the prism. This higher resolution revealed a series of dark lines that were explained in 1833 by William Miller of King's College London who had shown that gases absorbed light at specific wavelengths. The light from the sun was absorbed by gases present at its surface. This led to the use of the spectroscope to show that gases emitted light at specific wavelengths and that the same gases were most effective at absorbing light at those same wavelengths. In 1826 Fox Talbot, who later discovered practical photography, showed that heating substances in a Bunsen flame yielded a spectrum of wavelengths that would more readily identify the elements than wet chemistry - a remarkable shortcut to chemical analysis.

All these discoveries were produced by a small cadre of scientists, mostly men of means working for the sake of their intellectual curiosity. As science became more fashionable new revelations began appearing with more frequency.

The next milestone was photography, discovered simultaneously by Fox Talbot in England, and Niepce and Daguerre in France, the former's emulsion technique later proving to be more practical. The Fox Talbot story has its dramatic elements, particularly the legal aspects:

Lord Chief Justice Jervis presided over a famous trial at the London Guildhall from 18 to 20 December 1854. Plaintiff in the case was the already newsworthy scientist and discoverer of a technique for the photographic recording of images, William Henry Fox Talbot. Lined up as witnesses to support Fox Talbot's claim of infringement of his patent was an impressive list of experts including three distinguished professors of chemistry. The defendant, Martin Larouche, was a professional photographer who openly flaunted the Talbot patent and dared him to sue.

Public opinion, supported by several photography magazines, was on the side of the defendant and the court room was packed with spectators out for Talbot blood. Talbot, on the other hand, had engaged one of Europe's best lawyers, Sir William Grove, himself a scientist and Fellow of the Royal Society as well as *aficionado* of the ten-year-old art of photography.

Testimony centered on aspects of preparing the photographic paper, notably the collodian chemistry, but neither the judge nor jury fathomed the technical arguments. When Lord Chief Justice Jervis summed up the case for the jury it was clear they totally relied on the judge's limited understanding of chemistry. Talbot lost the decision, announced to a cheering court room, and this left him a defeated and disconsolate man. His initial urge to appeal the case gave way to acceptance, and license fees for the Talbot process were no longer collected. Talbot's years of effort proved a commercial loss. Even his later honors scarcely atoned for the anti-Talbot sentiment aroused in early Victorian society. He later wrote:

Nothing could be more illusory than such a trial. Neither judge or jury understood *anything* of photography. . .it was as if I or any other landsman were called upon to hear evidence and pronounce judgement, on the conduct of a naval captain in a gale of wind. In that case I should doubtless err greatly in my conclusions yet not so greatly as the Lord Chief Justice the other day. At any rate I should not confound the mainmast with the mizzenmast. I should know stem from stern. Would it be believed that *etiquette* prevented my counsel from interrupting the judge in summing up, and pointing out to him that he was going quite astray?

A Scottish academic at the time declared that Talbot would have received a fair trial in Scotland since the Scottish judges were well-versed in photography. Notwithstanding judicial incompetence in instructing the jury, the public had taken to this new diversion with a passion and it was hardly in the mood to buy a license. Talbot later granted amateurs a release from this obligation but public sentiment found few willing to acknowledge Talbot's claim by this gesture. His early rival in France, Daguerre, lost a similar legal fight over patents but the Talbot paper process rapidly outdistanced the Daguerre copper plate in practicality and the French technique faded from the scene.

Talbot was born on 11 February 1800. His mother, Lady Elisabeth Fox-Strangeways, was widowed when Henry was five months old and she married a Captain Charles Feilding (the spelling is correct). Henry demonstrated his precociousness before he was ten and was sent to Harrow in 1811. His notebooks at this time reveal a keen interest in science and languages and he appears to have got on well with his stepfather and schoolmates. Much of his correspondence of this period survives since he instructed his parents to preserve his letters.

In 1812 his chemistry experiments at Harrow resulted in an explosion and his tutor advised him to confine future work to theoretical matters. His mother wrote to him about this:

> . . .as he can no longer continue his experiments in Dr. Butler's house, he resorts to a good-natured blacksmith who lets him explode as much as he pleases.
>
> He has been trying to gild steel with a solution of nitromuriate of gold. This was the fatal experiment which blew him up at Dr. Butler's as it exploded with the noise of a pistol. . .Dr. B. was alarmed and declared that the Sun Fire Office would not ensure his house for a single day.

In one letter home he advised his parents that he was taught at school to be frugal and he immediately cut short his letter to demonstrate this. His brilliance at Harrow prompted the headmaster to ask him to leave school early at age 15, since he was considered too young to be head boy in his final semesters. His parents engaged a private tutor for this last year.

On leaving school Talbot visited Paris and then Waterloo, the latter only a year after the famous battle. He described the experience:

> A Flemish peasant was our guide, who had remained in his cottage during the battle - only in the cellar. Walked all over the field of battle, first to Hougomount, which is in ruins, the door pierced with musket holes in every part. The musket balls which were lodged in the trees in the greatest abundance have all been carefully cut out for sale, & the trees themselves cut down. The cannon balls have made great holes in the garden walls, & in the fruit trees that grew against them. One shot went through

two walls and killed seven men. Saw marks of blood on the wall. Only 25 English were killed in the garden, the French lost 600 in endeavouring to enter. . .The sabres, swords, helmets, cuirasses & other more valuable relics of the fray are now become scarce. Fresh arrivals of bullets and buttons from London to be sold on the field, warranted genuine. . .Returned to Brusells and dined again!!

When Talbot acquired a microscope he began a life-long study of botany. His ability with mathematics became evident when he made calculations of lunar eclipses and discovered errors in the published nautical almanac. In choosing between Oxford and Cambridge his mother opted for the former, feeling that Cambridge would place too much emphasis on mathematics. In a jovial vein she wrote to him to explain:

> You seem so mathematically inclined that I ought to send you to Oxford to counteract it, that you may not grow into a rhomboidal shape, walk elliptically, or go off in a tangent - all of which evils are imminent if you go to Cambridge.

He still preferred Cambridge and sat for the mathematics tripos exam at 21, a three day affair that included Euclid, algebra, trigonometry, mechanics, optics, astronomy, Newton's *Principia*, calculus, logic, moral philosophy and Christianity. The exam was held in unheated halls in January prompting Henry to comment, "The University had in its wisdom added corporal to mental fatigue!"

Henry won the University Greek prize in his first year and the classics prize the second. His explanations for second best in classics was contained in a letter to his stepfather in which he contended that he did not understand the results. He still thought himself better than the student who took first!

After finishing Cambridge in 1821 Henry took up residence at the family home in Lacock Abbey, Wiltshire. In 1822 he reduced the rents of the tenants because of a general depression and was thereafter considered a caring landlord. The income from rents provided him with some financial independence and he spent six of the next nine years in France, Switzerland, Belgium, Austria and Germany to study languages and bring back botanical samples. The problem of the Rosetta Stone caught his interest, as it did Thomas Young, and he amused himself with hieroglyphics.

Even before he knew how to record images permanently he experimented with the pinhole camera, using improved lenses that he acquired from Bavaria. His interests encompassed many fields: he suggested that falling stars were meteorites burning in the atmosphere; he pondered whether one could make ice by the evaporation of a volatile fluid in vacuum; he investigated the identification of materials by flame testing since most elements emitted distinctive colors when heated - in short he became the proverbial Renaissance Man. He even published a book of folk tales, only to be discouraged by his mother who wrote:

They are all very pretty. . . .but to form a volume of them
would I fear disappoint expectations so highly raised as they
have been about your talents, which have always justly been
held to be of a superior order. . . . If you had published anything
scientific, historical, or political. . . .but knowing as I do your
sterling abilities I do not like these tales and poems to be taken
for what you can do. . . .

His reputation as a good landlord spared him from the spate of riots that
visited England from 1830 to 1832 when farmers destroyed farm machinery,
mechanization being perceived as a threat to their jobs.

As Lord of the Manor, it is not surprising he sat for Parliament in 1832, the
same year he took a wife. The couple headed for Lake Como and it was here while
engaged in sketching that Henry thought of the desirability to record images
permanently, an idea that drove him to action three years later. His disillusionment
with Parliamentary affairs prompted him to step down after only a year with
the comment:

For my part I don't comprehend anything about it, nor who
can act with who, and why the remainder cannot. It is all an
enigma to me. . . .for many months the state of the Government
has been like a magazine of combustibles to which the hand
of a child might at any time have applied a match.

He turned his attention to optics when he noticed how a burst of light could
freeze a moving image, but it wasn't until 1835 that he first observed the darkening
of paper soaked in silver nitrate when exposed to light. Unfortunately he suffered
the temperament of the dilettante as his interest quickly shifted to mathematics,
winning the Royal Society medal in 1838. Years later his mother berated him for his
indecisiveness, pointing out that if he had not wavered in his work on photography he
would have beaten Daguerre.

As it was, Talbot was surprised to hear that the Frenchman had bested him in
capturing a permanent image on a metal plate. He rushed into action and improved
his process of saturating paper with silver nitrate. In January 1839 Faraday directed
people's attention to a display of Talbot's work at Rumford's Royal Institution. His
process was sufficiently different from Daguerre's that no one accused him of
imitation. While the Frenchman's technique gave photographs of superior clarity,
only Talbot's process permitted a positive photo and multiple print possibilities.
Talbot read his paper at a meeting of the Royal Society on 31 January 1839.

The technology required the discovery of hyposulphate of soda to fix the print
and render the silver nitrate benign. Talbot improved the speed of his film and in
1841 he secured a patent.

After receiving it Talbot sold licenses, inflaming the amateurs drawn to the
process. Until the advent of photography, the cost of purchasing a patent likeness

by an artist was prohibitive, hence the ardor felt by people who were asked to pay for the privilege.

Talbot was even criticized by artists who felt that their craft was in danger. Talbot rebutted prophetically as follows:

> . . .there is ample room for the exercise of skill and judgment. It would hardly be believed how different an effect is produced by a longer or shorter exposure to the light, and, also, by mere variations in the fixing process, by means of which almost any tint, cold or warm, may be thrown over the picture. . . .All this falls within the artist's province to combine and to regulate; and if. . . .he becomes a chemist and an optician, I feel confident that such an alliance of science and art will prove conducive to the improvement of both.
>
> We have sufficient authority in the Dutch school of art, for taking as subjects of representation scenes of daily and familiar occurrence. A painter's eye will often be arrested where ordinary people see nothing remarkable. A casual gleam of sunshine, or a shadow thrown across his path, a time-withered oak, or a moss-covered stone may awake a train of thoughts and feelings, and picturesque imaginings. . . .

Talbot's sensitized paper made him aware of the ultra-violet invisible component in the sun's spectrum that was capable of passing through a prism, as he reports:

> Experimenters have found that if the spectrum is thrown upon a sheet of sensitive paper, the violet end of it produces the principal effect: and, what is truly remarkable, a similar effect is produced by certain invisible rays, which lie beyond the violet, and beyond the limits of the spectrum, and whose existence is only revealed to us by this action which they exert. Now, I would propose to separate these invisible rays from the rest, by suffering them to pass into an adjoining apartment through an aperture in a wall or screen or partition. This apartment would thus become filled (we must not call it illuminated) with invisible rays, which might be scattered in all directions by a convex lens placed behind the aperture. If there were a number of persons in the room, no one would see the other: and yet nevertheless if a camera were so placed as to point in the direction in which any one were standing, it would take his portrait and reveal his actions.
>
> Alas! That this speculation is somewhat too refined to be introduced with effect into a modern novel or romance; for what a dénouement we would have, if we could suppose the secrets of the darkened chamber to be revealed by the testimony

of the imprinted paper.

One of Talbot's premonitions was realized in 1851 when he took a flash photograph of a moving object, employing a lamp supplied by Wheatstone. Talbot described its potential:

> Photographic portraits will be obtained with all the animation of full life instead of the stiffened serenity which even a sitting of a few seconds gives to the countenance. Nothing will be more easy than to take the most agile ballet dancer during her rapid movements, or to catch the image of the bird of swiftest flight during its passage. . . .

A fight later developed over Fox Talbot's contribution to photography. He was criticized by the Art Journal:

> Reviewing Mr. Fox Talbot's labours as an experimentalist, we find him industriously working upon the ground which others have opened up. . . .He has no claim to be considered as the discoverer of any photographic process, but merely as the deviser of processes from the results of other men's labours.

And in an exchange of letters in 1852 with Lord Rosse in *The Times*, Talbot made a generous offer to the nation:

> . . .It is very desirable that we should not be left behind by the nations of the Continent in the improvement and development of a purely British invention; and as you are the possessor of a patent right which will continue for some years, and which may perhaps be renewed, we beg to call your attention to the subject, and to inquire whether it may not be possible for you to obviate most of the difficulties which now appear to hinder the progress of the art in England. . . .

Fox Talbot replied:

> . . .after much consideration, I think that the best thing I can do, and most likely to stimulate further improvements in photography, would be to invite the emulation and competition of our artists and amateurs by relaxing the patent right, with the exception of the single point hereinafter mentioned, as a free present to the public. . . .

The exception was the employment of photography by commercial artists. Following the unsuccessful patent suit in 1854 Talbot turned his attention to other matters. The legal fees had been costly and he welcomed the sale to the Great

Western Railway of the right of way through Lacock Abbey, earning him £5000. He then devoted his time to recording the sun's eclipse.

Afterwards Talbot concentrated his effort on producing a printing process for photographic images. In both lithography and intaglio he succeeded in transferring his prints to ink.

One of the more intriguing ideas in the optical field was published by Talbot who studied the interaction of certain elements with water:

> The bright light given off under these circumstances by strontia, sodium, thallium, and many other substances, is very beautiful, and so permanent that at the close of the experiment the original grain of the substance does not appear diminished, and even the drop of water is found remaining unchanged. Provided always that the chemical substance is not one liable to decomposition under these circumstances of heat and moisture. . . .
>
> This method might be usefully applied to the illumination of microscopic objects by homogeneous light.

What a fascinating possibility to produce a light source driven electrically a good 20 years before Edison's incandescent bulb!

Talbot died in September 1877, a knighthood having eluded him. There is a museum at Lacock Abbey dedicated to his work.

Photographic imaging aided the field of astronomy by providing a permanent record for the relative position of heavenly bodies. Advances in telescope construction, notably the use of silvered glass to replace speculum metal as the reflector, provided better resolution in recording heavenly bodies. In 1838 a high performance telescope enabled Bessel in Königsberg to determine from parallax the distance to the star 61 Cygni as 600,000 times the distance from the sun to the earth.

In 1846 the planet Neptune was discovered after deviations in the orbit of Uranus were noted. Nebulae were observed in 1864, and in 1868 the Doppler shift toward the red of 30 miles per second was discovered for Sirius - optical technology was expanding exponentially. The next far-reaching and logical discovery came to the fore when Eadweard Muybridge entered the scene and invented the motion picture, although his successful scientific career made him good copy for the salacious details of his private life:

One evening in November 1874, Harry Larkyns, a dashing young ne'er-do-well was playing cribbage with friends in a cabin in Northern California, when one of the players responded to a knock at the door. The visitor asked for Mr. Larkyns and was invited in, but he indicated his business was private and he preferred to transact it on the porch. Larkyns was informed and left the game to attend to the stranger. After identifying himself in the dim light of evening, Larkyns had a pistol discharged into him, dying instantly. The murderer entered the cabin and, spying a few women said, 'I'm Eadweard Muybridge, please forgive this intrusion.' The trial that followed became a *cause celèbre* since Muybridge had already achieved fame as a photographer.

He was born on 9 April 1830 in Kingston-on-Thames, an old town near London. His father ran a corn and coal business but as soon as Eadweard was old enough he headed for California as the San Francisco agent for the London Printing Co. He arrived two years after gold was discovered and started a business selling lithographs of newsworthy events and Audubon drawings. In 1860 he turned his attention to photography, returning to England to purchase equipment and to make contacts in the business. On the journey east his stagecoach overturned, killing several passengers and leaving Muybridge with a head wound. After reaching London via New York and obtaining the necessary cameras, he returned to San Francisco via Panama, the same route taken by Michelson as a lad.

In 1867 he took some of the first pictures of Yosemite, publishing 260 views and achieving recognition in this medium. The US Government commissioned him to photograph Alaska shortly after it was purchased from Russia by Cabinet Secretary Seward. On 21 October 1868 San Francisco had a severe earthquake and Muybridge photographed the ruins. He invented the sky shutter, a device for limiting the exposure of the brighter sky in order to bring out details of the darker land masses. Other commissions included photographing the US Cavalry in action and recording scenes of railroad rights of way. Turning to stereo photography, Muybridge became recognized as the leading photographer of the San Francisco area.

In 1872 he was engaged by California Governor Leland Stanford to photograph his family, and a working friendship endured for the next decade, until a misunderstanding caused a permanent rift. Stanford was a horse fancier and made a bet that all four legs of a trotter never left the ground. With $25,000 at stake Muybridge was engaged to settle the bet. He developed a fast shutter to prove Stanford's assumption.

At 40 he married Flora Stone, a girl half his age, and a few years later discovered that their first son was sired by Harry Larkyns. After committing the murder, Muybridge was held in the Napa jail awaiting trial. He was indicted in 1874 and, almost immediately, Flora sued for divorce on grounds of cruelty. Since everyone knew the circumstances surrounding the case, the divorce petition was dismissed.

The trial in February 1875 was nationally reported. Muybridge's defense was justifiable homicide since California law did not recognize jealousy as a motive. During the trial his attorney tried to show that Muybridge's stage coach accident caused him to have fits of temporary insanity. One witness even claimed that Muybridge must have been insane standing at the precipice of Yosemite to be photographed. When Muybridge took the stand he related details of the stage coach accident but nothing about the actual Larkyns' killing. The jury deliberated 13 hours and acquitted him.

Recovering from the trauma of the trial, Muybridge returned to work and produced a unique series of panoramic views of San Francisco in 1876 as well as continuing with his work on the moving horse. In collaboration with Stanford, who funded the $42,000 effort, Muybridge employed a bank of eight cameras to produce a sequential series of photos with electromagnetically triggered shutters. To view the moving series of pictures, the zoöpraxiscope was developed employing a rotating wheel with time sequential photos and a lens for projection. This predated Edison's motion pictures by ten years.

Leland Stanford delivered a report of the work on a visit to Paris in 1881 and this was followed shortly afterwards by one by Muybridge himself. Before such dignitaries as the Prince of Wales, Tennyson, Gladstone, Tyndall and Huxley, Eadweard Muybridge delivered a paper on his work in the lecture hall at Rumford's Royal Institution in London, and the event was duly recorded in the *Photographic News*:

MR. MUYBRIDGE AT THE ROYAL INSTITUTION

Before a distinguished audience, which included H.R.H. the Prince of Wales, the Princess of Wales, and the three young Princesses, the Duke of Edinburgh - a distinguished photographer, it may be remembered - the Poet Laureate, the President of the Royal Society, and most of the managing body of the Royal Institution, Mr. Muybridge, of San Francisco gave on Monday his first public demonstration in the country. Mr. Muybridge may well be proud of the reception accorded him, for it would have been difficult to add to the éclat of such a first appearance, and throughout his lecture he was welcomed by a warmth that was as hearty as it was spontaneous.

Mr. Muybridge wisely left his wonderful pictures to speak for him, instead of making the occasion the subject of a long oration. He showed his photographs one after another on the screen by the aid of an electric lantern, and modestly explained them in clear but plain language. In this way the demonstration was at once rendered entertaining as well as interesting.

Mr. Muybridge first explained his plan of securing such rapid pictures of animals in motion. He showed a representation of his 'studio' to begin with; it was like that portion of a racecourse to be found opposite the grandstand. This latter building was, in effect, a camera stand, and a very good one into the bargain, for it contained twenty four cameras in a row, the lenses a foot apart, all looking on to the course. As the animal passed, these cameras, with their instantaneous shutters, were fired off one after another with electricity. Thin linen threads, breast high, and a foot apart, were stretched across the course, and as the animal broke these threads, they, being connected each of them with a camera, brought about the exposure. The instantaneous shutter in each case simply consisted of two little planks, one to move upwards and the other to move downwards, in front of the lens by rubber springs; the tension in these latter - equal to 100 lb., Mr. Muybridge said - and the exposure was calculated to be $1/5000$ of a second. Whether this calculation is correct or no, certain it is that the spokes of a trotting carriage shown were very sharp, and there is hardly a movement visible in any of the animal pictures.

We may mention here that all photographs were taken on wet plates, for they were secured four years ago. Iron was employed in their development, and no additional care or particular method was had recourse to.

Mr. Muybridge, by way of comparison, first threw on the screen a series of artist's sketches of the horse in motion, some of them old world designs of the Egyptians or Greeks, some very modern, including the principal animal from Rosa Bonheur's well-known 'Horse Fair'. In no single instance had he been able to discover a correct drawing of the horse in motion and, to prove his statement, he then threw on the screen several series of pictures representing the different positions taken up by a horse as he walks, trots, ambles, canters, or gallops. One thing was very plain from Mr. Muybridge's pictures, namely, that when a horse has two of his feet suspended between two supporting feet, the suspended feet are invariably lateral; that is to say, both suspended feet are on the same side of the animal. This, no painter - ancient or modern - had ever discovered. Then the amble was found to be different from the canter, and the canter very different again from the gallop; although most people imagined that, to perform all these, the horse used his legs in the same fashion. Mr. Muybridge was at some difficulty to describe the amble, and it seemed at one time as if it would be necessary to call upon Mr. Tennyson to give a definition of it in his well-known lines: 'Property, property, property!' - but he succeeded subsequently in defining the step very satisfactorily afterwards by means of his pictures.

After Mr. Muybridge had shown his audience the quaint and (apparently) impossible positions that the horse assumes in his different gaits, he then most ingeniously combined the pictures on the screen, showing them one after another so rapidly, that the audience had before them the galloping horse, the trotting horse, etc. Nay, Mr. Muybridge, by means of his zöepractiscope showed the horse taking a hurdle - how it lifted itself for the spring; and how it lightly dropped upon its feet again. This pleasing display was the essence of life and reality. A new world of sights and wonders was, indeed, opened by photography, which was not less astounding because it was truth itself.

After these life-like pictures, it needed not Mr. Muybridge's dictum that to use a mild term it was "absurd" to see a galloping horse depicted with all four feet off the ground, a simple impossibility. And if this held good of one horse, what must be said of ten horses, thus painted, as was the case in Frith's 'Derby Day', which Mr. Muybridge projected on the screen by way of comparison, and by which the photographer described as a miracle.

Mr. Muybridge modestly calls his series of animals in motion - they include horse, dog, deer, bull, pig, etc. - simply preliminary results. They contain little or no half-tone, and are only proof of what may be done. What he desires now to secure, if he only receives sufficient encouragement, is a series of photographic 'pictures', and these, with the experience he has now acquired, and with the gelatine press to help him, should be well within his reach. We only trust this encouragement will be forthcoming, and that Mr. Muybridge will be tempted to carry on the difficult work he has commenced with such genuine success.

"I should like to see your boxing pictures," said the Prince of Wales to Mr. Muybridge.

"I shall be very happy to show them, your Royal Highness," responded the clever photographer, "I don't know that these pictures teach us anything useful, but they are generally found amusing."

Stanford and Muybridge prepared a joint book on their efforts. Stanford was dealing with his Boston publishers and inadvertently forgot to send Muybridge the proofs. When Eadweard was presented with the final publication, he was distressed that original photos were not used. He sued the publisher and Stanford for ruining his reputation. He also claimed that the book 'Horses in Motion' infringed his photo copyrights. Muybridge lost both suits and the subsequent bitterness caused him to leave California and transfer his allegiance to the University of Pennsylvania where the facilities of their veterinary department were put at his disposal. With a battery of 12 cameras and a series of fast shutters he recorded action shots of everything imaginable.

His fame spread and he responded to speaking engagements at the Royal Society, and numerous schools in England and Germany. In 1893 he had a special exhibit at the Chicago World's Fair.

Having never changed his citizenship, he settled down at Hampton Wick in Surrey, England, not far from his birthplace and published several further books on animals in motion.

He died on 8 May 1904. A permanent exhibit of his work is on display in Kingston-on-Thames.

While the study of light has enriched man's life in the great works or art, another technique that has already been mentioned is the cartoon, an amusing way to incorporate political satire within the economy of a single drawing. This was taken further by the team of Lewis Carroll and Punch magazine cartoonist John Tenniel, who provided us with the 1862 story of Alice in Wonderland and its instantly recognized images of characters like Alice, Tweedledum and Tweedledee, the Cheshire Cat, and the Mad Hatter.

Lewis Carroll successfully combined the life of a mathematician with that of the most widely read children's storyteller. His language and Tenniel's drawings provide visual images par excellence. The cartoonist draws political figures that are instantly recognized; Lewis Carroll and Tenniel provided us with figures that are as readily identified as Washington, Franklin, and Lincoln.

On 4 July 1862, while the Americans took a brief respite from their Civil War to celebrate the ninety-sixth anniversary of the Declaration of Independence, a significant event occurred on the peaceful Isis River near Oxford, England. The Reverend Charles Luttwidge Dodgson, mathematics don at Christ Church College, took the three daughters of the Dean on a boating trip and picnic. That event had a significant impact on the English-speaking world for it led to *Alice in Wonderland*. On that peaceful river outing the Reverend Dodgson related the impromptu story of Alice's adventures underground, appealing to the ten-year-old real Alice and her two sisters.

Since its publication *Alice in Wonderland* has become one of the most quoted books in the English language and the illustrations of Sir John Tenniel probably more familiar than the story itself. The combination of drawing and word imagery brings instant recall to most people - no study of light can overlook this. Less known was Dodgson's ability as a Victorian photographer, at a time when the technology was cumbersome and its use relatively non-artistic. Dodgson's photographs of little girls, particularly Alice, are gems of composition.

Only Alice's subsequent nagging of Mr. Dodgson eventually led to publication of Alice in Wonderland. Dodgson first tried his own hand at the illustrations but, recognizing his shortcomings, prevailed on 'Punch' cartoonist Tenniel to do the famous drawings. Since that time over 50 artists have published scenes from the story.

Alice in Wonderland emerged as the product of an imaginative mind stimulated by mathematical propositions and prolific letter writing. Dodgson kept a log of his letters and produced over 100,000 in his lifetime. This mixture of mathematics and English required the catalyst of the real-life Alice, daughter of Dean Liddell (pronounced Liddle), who with her sisters was treated to the story. Dodgson published the story under the *nom de plume* 'Lewis Carroll'.

There are two visual concepts in mathematics, symmetry and transformation, ever present in Carroll's writings. Symmetry refers to the way things can be rotated and made to repeat, like a three leaf clover every 120 degrees or a square every 90 degrees. Transformation makes things bigger or smaller, as befell Alice in exploring *Wonderland*, or shifting a repeatable pattern such as on a chessboard.

To the mathematician the application of these concepts becomes obvious in studying the events of *Wonderland*, but it is the employment of these concepts in the language that accounts for much of its popularity. Symmetry appears in the frequent use of *double entendre* or in words that sound alike but have multiple meanings.

In falling down the rabbit hole Alice considers the prospect of going through the entire earth and emerging where people walk on their heads 'The antipathies, I think' she utters as a pun on antipodes, almost immediately followed by Alice's thoughts as the whether cats eat bats, a symmetry in rhyme. Consider the lines written to Mrs. Emmie Drury:

"I'm EMInent in RHYme!" she said
"I make WRY mouths of RYE-meal gruel."

In which Carroll has incorporated portions of the recipient's name. The following limerick is based on *double entendre:*

There was a young lady of station,
"I love man," was her sole explanation.
But when men cried "You flatter,"
She replied, "Oh! no matter,
Isle of Man is the true explanation."

The following limericks employ dimensional transformations:

There once was a man of Oporta
Who daily grew shorter and shorter
The reason he said
Was the hod on his head
Which was filled with the heaviest mortar.

His sister named Lucy O'Finner
Grew constantly thinner and thinner
The reason was plain
She slept out in the rain
And was never allowed any dinner.

Another use of verse and transformation is the acrostic where the first letters form the word Margaret:

Maidens, if a maid you meet
Always free from pout and pet
Ready smile and temper sweet
Greet my lttle Margaret.
And if loved by all she be
Rightly, not a pampered pet,
Easily you then may see
T'is my little Margaret.

Carroll's humor and logic reveal his mathematics training:

Statements
1. No ducks waltz
2. No officers decline to waltz
3. All my poultry are ducks

Conclusion
My poultry are definitely not officers!

Carroll also concocted a game whereby words were transformed by changing a single letter. Change head to tail:

HEAD- heal- teal- tell- tall- TAIL

In his official role as curator at Christ Church he produced two reports:

Twelve months as a curator
By one who has tried it.

Three years as a curator
By one whom it has tried.

In another report overlooking the subjects of ventilation, lighting, and seating in the Common Room, he headed the section:

OF AIR, GLARES, AND CHAIRS

In Jabberwocky there is the symmetry of sounds in nonsense words:

And, as in uffish thought he stood
The Jabberwock, with eyes of flame,
Came whiffling through the tulgey wood
And burbled as it came.

In the original *Looking Glass* manuscript, Carroll requested that the entire poem be printed in its mirror image but, in sympathy for the typesetter, relented for the first verse alone.

The lion and the unicorn in *Looking Glass* had symbolic meanings, the lion for England and the unicorn for Scotland, as well as for Gladstone and Disraeli respectively.

This letter to his sister describes how he employs transformation in his lectures:

My dear Henrietta,
. . . .the tutor should be dignified and at a distance from the pupil, and the pupil as much as possible downgraded. I sit at the further end

of the room; outside the closed door sits the scout; outside the outer door (also shut) sits the subscout; halfway down the stairs sits the sub-sub-scout; and down in the yard sits the pupil. The questions are shouted one to the other:

Tutor: What is twice three?
Scout: What's a rice tree?
Sub-scout: When is ice free?
Sub-sub-scout: What's a nice fee?
Pupil (timidly): Half a guinea?
Sub-sub-scout: Can't forge any
Sub-scout: Ho for Jenny
Scout: Don't be a ninny!
Tutor: (offended): ?

In writing to his niece he admonishes her since, whenever he sends 'love' to her, she always sends it back as though she didn't value it! Or writing to Edith Jebb he says:

'Edith my dear! My cup is MT. Will you B so kind as 2 fill it with T?'

Or this bit or asymmetry in his letter to Margaret Cunnynghame:

You say that I'm "to write a verse" -
O Maggie, put it quite
The other way, and kindly say
That I'm "averse to write."

In his letter to Beatrice Hatch he says:

'You've no idea how careful we have to be, we dolls. Why there was a sister of mine who sat too close to the fire and one of her hands dropped off. It dropped right off because it was the right hand and the other was left.

The following problem and solution reflect a symmetry concept:

Problem: A and B began the year with only £1000. They borrowed nought and stole nought. On next New Year's Day they had over £60,000 between them. How did they do it?

Answer: They went to the bank of England. A stood in front and B stood behind.

The DODO bird is symbolic of the stuttering Charles Dodgson--Do-Do-Dodgson. The disappearance of the Cheshire cat, leaving only the grin, prompts the asymmetric remark by Alice: 'I've often seen a cat without a grin but never a grin without a cat.'

Alice in Wonderland and *Through the Looking Glass* represent about one fifth of Carroll's collected works (letters excepted). Even at 13 Carroll was imbued with mathematical logic:

> MY FAIRY
> I have a fairy by my side
> Which says "I must not sleep,"
> When once in pain I loudly cried
> It said "You must not weep."
>
> If, full of mirth, I smile and grin
> It says "You must not laugh,"
> When once I wished to drink some gin
> It said "You must not quaff."
>
> When once a meal I wished to taste
> It said "You must not bite,"
> When to the wars I went in haste
> It said "You must not fight."
>
> "What may I do?" at length I cried
> Tired of the painful task
> The fairy quietly replied,
> And said "You must not ask."
>
> Moral- "You mustn't!"

In *Sylvie and Bruno* we have another example of symmetry:

> He thought he saw a Banker's Clerk
> Descending from a bus
> He looked again and found it was
> A hippopotamus
> "If this should stay to dine," he said,
> "There won't be much for us!"

Carroll carried this into his sequel *Sylvie and Bruno Concluded*:

> He thought he saw an argument
> That proved he was the Pope

He looked again, and found it was
A bar of mottled soap,
"A fact so dread," he faintly said
"Extinguishes all hope!"

Dodgson's personal life remains an enigma. His stuttering, except with little girls, and his asexuality may be related to his Spartan training when he was sent to school in Yorkshire. He was taunted and teased and was an easy prey for pranks by the other boys. He remained a bachelor and one wonders if he had some homosexual proclivities. But, if so, why did he delight in photographing little girls, *PARTICULARLY IN THE ALTOGETHER?*. The interest was probably not prurient. He has stated that he always found a pristine beauty in unclothed little girls. He may even have been born with atrophied genitals but, whatever the explanation, the visual imagery of his works remains one of the highlights in the use of light to enrich our lives. His characters from *Alice in Wonderland* are as well known as the Mona Lisa, Tower Bridge, Statue of Liberty, and the Eiffel Tower.

A contemporary of Lewis Carroll, although we have no evidence that they met, was the King's College London physics professor who invented the concertina, Charles Wheatstone. They would have enjoyed each other's company since Carroll was attracted to math puzzles and Wheatstone invented a system for encoding messages that was adopted by the Admiralty.

On a Friday in April 1846, Wheatstone was scheduled to give one of the popular science discourses at Rumford's Royal Institution in Albermarle Street. The attendees, dressed formally for these auspicious evening occasions, were all seated at the appointed hour when the Director, Michael Faraday, entered, gave some excuse for the absence of the lecturer, and delivered an impromptu lecture instead. What Faraday failed to mention was that Wheatstone bolted at the last moment and ran down the stairs into Albermarle Street rather than face the audience. Since that day Friday speakers are locked in the anteroom one hour prior to their lecture.

The story has no doubt been embellished over the years, although the custom of locking in speakers appears to be true. It is certainly known that the brilliant Charles Wheatstone could produce lucid explanations in private, but he quaked in his boots before formal audiences. This is not rare amongst scientists.

Charles Wheatstone was born on 6 February 1802 in Gloucester, England, but the family moved to London in 1806 where the father set up shop as a maker of musical instruments. He evidently attained some distinction since he subsequently gave music lessons to Princess Charlotte.

Charles was a genius with an independent streak. He could read at three and often disagreed with his school teachers in Kennington, even prompting him on one occasion to run away. He reached Windsor, over twenty miles away, before being tracked down. When informed later that he had won the school's French Prize, Charles turned it down rather than give the mandatory acceptance speech. He displayed early talent in mathematics, physics, songwriting, reading and constructing models. To a great extent, though, he was self-taught.

The environment of the music shop prompted him to take an interest in Chladni patterns. Some years previously the German physicist Chladni had demonstrated that fine sand sprinkled on the soundboard of an instrument would move into interesting visual patterns when the instrument was sounded. Those parts of the soundboards that vibrated strongly would shake the sand and cause the grains to move towards regions with smaller amplitude. Wheatstone improved on these patterns by employing finer powders such as ground talc and extended the investigation by submerging the soundboard just below the water level where surface waves could be excited. Today laser excitation of soundboards yields even clearer standing wave patterns, aesthetically interesting but not yet capable of identifying a fine instrument. The psycho-physiology of the ear is acute but distinctly different from the eye.

Wheatstone extended his interest to the transmission of sound in solids. He may have 'froze' as a speaker before large audiences but he developed a penchant as a showman. He devised the Enchanted Lyre or Acoucryptophone (hearing a hidden sound). A large lyre was suspended by a wire that passed through a small hole in the ceiling and was attached to a musical instrument on the floor above. When the instrument above, such as a piano, was played the vibrations passed along the wire and set the lyre into resonant vibrations. To the audience seated below it defied explanation. The Enchanted Lyre was taken 'on the road' and proved a commercial success, receiving favorable reviews from critics.

When Charles was 21 he and his brother inherited the music business but this did not deter his continued interest in all aspects of physics. He developed a device to measure persistence of vision, an opaque disc with a pie-shaped sector removed. By spinning the disc and observing a scene behind it he could relate the minimum speed necessary to produce an unbroken image. The typical persistence time of 0.1 second blends the individual frames of a movie into a continuous picture. It accounts for the time of traversal of the current of ions from the eye to the brain plus the time required for the brain to respond. It also corresponds to the time required for the eye's pupil to dilate under the action of a sudden change in light intensity and the time needed for the eye and brain to recognize an object or a face.

Wheatstone devised a stethoscope to detect vibrations when it was placed in contact with an instrument's soundboard, an interesting variant of Chladni patterns. He demonstrated the onset of turbulence in a vessel of boiling liquid by listening to the vibrations transmitted to the vessel's outer glass walls. He studied the resonances in the mouth's cavity during the playing of the Jew's harp, elevating the simple instrument into the performance class. A man was found who could form his mouth so as to excite two resonances and whistle a duet. Wheatstone worked on a technique for sending messages over long distances through solid rods but discovered that the rods had to be extremely homogeneous to prevent scatter and loss of sound. While this work is considered *passé* today, one contribution is still in fashion, the concertina. In Wheatstone's own convoluted verbiage:

. . .the employment of two parallel rows of finger studs on each end or side of the instruments fitted with keys to terminate the ends of the levers of the keys, and then so placing them with respect to their distances and positions as that they may, singly, be progressively and alternately touched or pressed down by their first and second fingers of each hand, without the fingers interfering with the adjacent studs, and yet be placed so near together as that any two adjacent studs may be simultaneously pressed down, when required, by the same finger, the peculiarity and novelty of this arrangement consisting in this, that as the ordinary keyed wind musical instruments the fingering is effected by the motion sideways of the hands and fingers, in this new arrangement that mode of fingering is rendered entirely inapplicable, and a motion which had not hitherto been employed is rendered available, namely, the ascending and descending motions of the fingers before described. This mode of arranging the studs enables me to bring the keys much nearer together than has hitherto been done in any other instrument of a similar nature, and thereby to construct such instruments of greater portability.

In 1833 Wheatstone and Fox Talbot crossed paths, the latter not yet involved in photography. Talbot had been awarded a Fellowship of the Royal Society in 1831 and was serving as a Member of Parliament when he published a critique on Wheatstone's determination of the velocity of electricity. Wheatstone acknowledged some appropriate corrections in a follow-up letter. His later values for the speed of electricity agreed with the values for the speed of light that had been determined by observing the eclipses of the moons of Jupiter. Wheatstone suggested a rotating mirror technique to determine the speed of electric pulses, a method taken up 8 years later and applied to light by Foucault.

King's College London was opened in 1829 and five years later Wheatstone was appointed Professor of Experimental Philosophy (physics). Since the duties of the post were not specifically detailed Wheatstone took some liberty in teaching his students. He allowed them access to his music shop where there were machines to fabricate instruments. Wheatstone remained at this post until his death in 1875. During those years his interest centered on electricity, its production, measurement and practical applications. He and Fox Talbot worked on the linear motor, used to drive a reciprocating engine.

He developed the Wheatstone Bridge for measuring electrical resistance, worked on a device to electrically detonate explosives underwater and clear the Thames of wrecks, produced a sensitive galvanometer to measure electric charge and, finally, turned his attention to the telegraph.

In partnership with William Cooke patents were secured and a profitable enterprise developed as the telegraph underwent gradual improvement. Oddly enough, when Wheatstone suggested that the telegraph might effectively replace the semaphore stations for visual communication between London and the Portsmouth Naval Base, the Admiralty turned him down.

The partnership with Cooke had some rough sailing but profitability calmed many of their disputes. During this period Samuel F.B. Morse arrived from America seeking a British patent on his own version of the telegraph but having published details in a magazine the patent applications became invalid in England. To this day America credits Morse as the inventor while Wheatstone is so honored in England.

Between 1836 and 1844 Cooke and Wheatstone made hundreds of thousands of pounds out of the telegraph prompting Wheatstone to turn his attention to the underwater telegraph. For this he recognized the need for insulated cable and experimented with tar surrounding the conductor inside a metal pipe. The interest focused on a cross channel telegraph.

In June 1843 Prince Albert visited Wheatstone at King's College to witness a demonstration of the underwater telegraph by having a message sent through a pair of cables strung across the Thames. While the experiment ran into some difficulty and had to be postponed, Wheatstone later received a message from the Prince asking why two cables were necessary - could not the return signal be transmitted by the water itself? This display of erudition was impressive for Royalty. Wheatstone's reply has never been unearthed.

In his legal dispute with Cooke, Wheatstone gave his own recollection of the origin of the telegraph, an effort independently conceived by Samuel F.B. Morse in America:

> When I made in 1821 the discovery that sounds of all kinds might be transmitted perfectly and powerfully through solid wires and rods, and might be reproduced in distant places, I thought that I had an efficient and economical means of establishing a telegraphic (or rather a telephonic) communication between two distant places. Experiments on a larger scale, however, showed me, that though on short distances most perfect results were obtained, yet that the sounds could not be efficiently transmitted through very long lengths of wire. . . .
>
> I afterwards turned my attention to the employment of electricity as the communicating agent. . .if the velocity could be proved to be very great, there would be encouragement to proceed. I. . .ascertained that electricity of high tension travelled through copper wire with a rapidity not inferior to that of light through planetary space, and I obtained abundant reasons for believing that electricity of all degrees of tension travelled with the same velocity in the same medium. . . .
>
> In the year 1836 I repeated these experiments with several miles of insulated wire,. . .I devised a variety of instruments by which telegraphic communications should be realized on these principles.

Years after their 1840 differences Cooke published his side of the disagreement:

> The invention became a subject of public interest; and I found that Mr. Wheatstone was talking about it everywhere in the first person singular. . . .At length, in 1840, I required that our positions should be ascertained by arbitration.

Arbitration and a liberal sprinkling of profits settled the case.

Wheatstone was married on 12 February 1847 in a civil ceremony that preceded the birth of his child by three months. *The Times* did not report the marriage between Professor Charles Wheatstone and Emma West for reasons obvious in Victorian times although it did report Wheatstone's attendance at a Royal Society meeting! The couple produced five children but little else is known of the relationship.

At this stage in his career the Crimean War captured the attention of the nation. In particular, the telegraph heightened interest in ciphers and codes. Wheatstone developed his cryptograph which permitted encoding by a key word. This technique was employed by the British in the Boer War and 60 years later in the Great War.

During his period of fame as a cryptanalyst the British Museum sent Wheatstone a three-page message originally transmitted in a numerical code by Charles I 200 years previously. After finally 'cracking' the message Wheatstone apologized for the delay - the Museum had failed to tell him the message was in French.

Wheatstone devoted some time to the stereoscope. He recognized that two eyes observed slightly different perspectives and he made a series of drawings to demonstrate the effect. A card held vertically between the drawings helped the eyes in making the separation.

With the advent of photography Wheatstone induced Fox Talbot to provide him with stereoscopic pairs of his photographs to demonstrate the effect. Wheatstone also devised a clock that recorded the time of day depending on the diurnal variation of the polarization of sunlight. He dabbled with the spectrograph and showed that a spark drawn to a metal had distinct lines when viewed through a prism.

He was knighted on 30 January 1868. King's College has established a Wheatstone Chair in physics and displays some of his inventions in the department's museum. His achievements should have produced a man of unlimited confidence, yet he trembled on stage. It should be added that it is not a requirement for the incumbent Wheatstone Professor to be a poor speaker. How ironic that both Lewis Carroll and Charles Wheatstone had significant speech difficulties!

The physics of light proceeded through a series of experiments in the nineteenth century that culminated in the accurate measurement of the velocity of light by Michelson and the magnificent theoretical work of James Clerk Maxwell.

The approximate speed of light had already been measured by Roemer from the delay time in the orbit of the moons of Jupiter depending on the added distance across the earth's orbit around the sun. Roemer's measurement was in error by only forty percent. In the nineteenth century Foucault had performed a simple test to make this determination. A stationary mirror reflected a fine slit of light to a distant mirror and the position noted. The mirror was then set into high speed angular rotation and the shift of position noted. By the time the light had travelled the distance and back, the mirror had changed angle and the light is reflected to a different position. Typically, for a distance from the light source to the mirror of 20 meters and an angular velocity of 417 rps the light beam will move about 0.7 mm. By placing a tank of water in the light path the velocity was shown to decrease and the index of refraction (ratio of the speed of light in air to the speed in water) determined.

The concept of the aether as the medium responsible for the conduction of light was an accepted part of classical physics, just as air was recognized as the transmitting medium for the conduction of sound. The aether dates back to ancient times and Michelson's attempt to detect it was to provide an accurate value for the relative motion of the earth within the aether. The last thing Michelson and his co-worker Morley expected was a null answer. The result was a bitter disappointment.

Michelson made his first determination on a visit to the Helmholtz laboratory in Berlin (1882), but it hardly caused a stir, although two mathematicians, Fitzgerald and Lorentz, suggested years later that matter shrinks in compensation as it approaches the velocity of light, thus yielding an apparent zero effect. When Einstein's theory of relativity later explained the absence of an aether, physicists took note. Michelson, himself, did not accept Einstein's explanation - he still thought as Newton did:

> To suppose that one body may act on another at a distance
> through a vacuum without the mediation of anything else, is
> to me so great an absurdity that I believe no man, who has in
> physics a complete faculty for thinking, can ever do.

Historians of physics generally agree that Einstein was not significantly influenced by the Michelson-Morley experiment as he developed his ideas about relativity. Nevertheless, his theory produced a brilliant resolution of the puzzling zero aether drift and showed how matter behaved as it approached the velocity of light.

Michelson, like Thomas Edison, was a painstaking experimenter whose sense of the physical world depended on what he could see and feel. Deep theoretical thinking was left to others. His accurate determination of the velocity of light, his observation of a null aether drift, his measurement of the length of the standard meter bar, and his determination of the size of a star rank him as one of the foremost contributors in optics. His humble beginnings added to the mystique.

Michelson was born into a Jewish family in Poland on 19 December 1852 at a time when the Jews were being subjected to discrimination in Eastern Europe and many emigrated to America. In most cases, as with Michelson's father, they had no assurance of employment, but as a dry goods merchant he invariably found customers everywhere. No sooner had the Michelsons arrived in New York with two-year old Albert, than the decision was made to try their luck in goldrush California. The family sailed to Panama, trekked overland through a disease-ridden country and then took a ship to San Francisco. The family set up shop in Calaveras County, site of the famous Jumping Frog story of Mark Twain.

What a rough and tumble place for a lad who was destined to become a scientist! Life was not influenced by aesthetics or learning, yet young Albert managed to find someone to teach him the violin and to discover he had a flair for sketching. At 13 he entered high school in San Francisco but the rapid exhaustion of the gold fields prompted the family's move to Virginia City, Nevada, where the famous Comstock Lode silver fields had been discovered. The boy's aptitude for learning was apparent to his parents and a solution was sought for his post-high school education, since the family could not afford college. Albert sat for the US Naval Academy exams for the Virginia City district. He tied with the son of a disabled Civil War veteran who was given preference. Not discouraged and armed with a letter of recommendation, the lad of 16 travelled unaccompanied to Washington to seek one of the ten special presidential appointments.

Albert demonstrated his resourcefulness by intercepting President Grant while he was walking his dog. He presented his letter of recommendation but was informed that all ten appointments were taken. Grant suggested that Michelson travel on to Annapolis; he could replace any of his ten appointees who failed the physical. Still hopeful, the lad journeyed to the Naval Academy, but he found that all ten appointees had qualified and he returned to Washington. The President reconsidered the situation and granted Michelson an illegal eleventh appointment.

No doubt the President was pressured to act. During the Civil War he had experienced an embarrassing situation in respect to the Jews (the polite term in those days was 'of Israelite Persuasion') when some dissatisfaction with a Jewish peddlar arose and, as General, he forbade all trade with Jews. President Lincoln suspended this order but Grant carried the stigma into the presidency. There were so few Jews at Annapolis that it was considered politically expedient to accommodate Michelson.

Michelson proved to be a good athlete and boxer at the Academy and he graduated ninth in a class of 20. During Michelson's lifetime, as with Einstein, Born, and Disraeli, he never practiced organized Judaism, but as Disraeli pointed out, he was always reminded of his heritage. Michelson stayed on at the school as an instructor in physics and chemistry and at 25 married an Admiral's niece.

With his not-too-pressing duties at Annapolis, he found time to make his first measurement of the velocity of light employing the Foucault rotating mirror method. In 1878 Michelson's determination was within a few tenths of a per cent of the presently accepted value.

Michelson now had caught the 'bug' and the minutiae in performing an accurate measurement drove him until he died 53 years later. Like the man who climbs mountains because they are there, Michelson confessed that his obsession was fun.

He secured leave to study and work abroad and spent time at the leading schools in Berlin, Heidelberg and Paris. While abroad he applied for the position of Professor at Annapolis.

During his stay with Helmholtz in Berlin he became infused with the more profound matter of the aether, a concept no one doubted. The measurement was simple; using a half-silvered mirror and an interference technique between two beams travelling at 90 degrees to each other, any motion of the earth relative to a universal aether would produce a phase shift. The Berlin measurements failed to uncover any effect and this null result was repeated in Potsdam close to Berlin where traffic vibrations were diminished.

When one is looking for something that one is convinced must be present and fails to find it, a man of Michelson's determination must have been driven mad! Nonetheless, he was sufficiently confident of his null result to publish in the French Journal, *Comptes Rendus*. No doubt Michelson considered two other possibilities: the effect was smaller than one thought, or there was some fault in the technique.

With the professorship at Annapolis not forthcoming, Michelson accepted one at Case Institute in Cleveland, and it was here that he met Morley, a chemist from Western Reserve University. Fifteen years his senior, and a graduate from theological school who could not find a vacancy in the pulpit, Morley's appearance contrasted sharply with the dapper Annapolis graduate. Morley also became infused with the measurement of the aether drift, and as an organist may have found himself engaged in violin duets with Michelson. The two complemented each other and decided to join forces.

Their first effort was an unsuccessful search for the aether drift in moving water. They then constructed a large apparatus with the optical equipment mounted on a huge stone floating in a bed of mercury, permitting easy rotation relative to the aether. After several years, they failed to uncover what nature was denying them, and they published this in the summer of 1887. Michelson must have engaged in deep soul-searching, having to face the raised eyebrows of his colleagues who wanted to know why he couldn't find a positive effect. He was not a gregarious man and preferred to eat alone, possibly to avoid the queries.

A glimmer of relief came from the two European mathematicians, G.F. Fitzgerald from Dublin and H.A. Lorentz from Leiden, who suggested that at speeds approaching the velocity of light a measuring stick contracts. To Michelson this hardly meant anything, since it was not something he understood within his world of mirrors and interferometers. As J. Willard Gibbs, the famous thermodynamicist, stated at the time, 'A mathematician can say anything, a physicist must remain partially sane.'

For an Einstein born in 1879, the Michelson-Morley experiment and the Fitzgerald-Lorentz contraction of 1895 perhaps came to his attention during his years reading physics journals as a patent clerk in Bern, Switzerland. While his ideas did not gel until the turn of the century, the aethereal question probably entered his subconscious. With the publication of Einstein's theory of relativity in 1905, physicists had an explanation for the null result - light did not require a transitting medium like sound. The Fitzgerald-Lorentz contraction became part of a theory in which both space and time changed with velocity. Most important was the simple understanding that the speed of light was a constant, no matter what the relative motion of the source and observer. Such motion only added or subtracted apparent energy from the light by changing its observed color, but this explanation never satisfied Michelson. In 1894 he accepted a new appointment and transferred his work to the recently completed physics building endowed by Rockefeller at the University of Chicago.

After this move Michelson turned his attention to the standard platinum meter bar reposing in Sèvres, France. While one could have access to a second bar calibrated against the original, one still had to correct for temperature and pressure and, worst of all, inaccessibility. By employing the 643.8 nm line from cadmium gas Michelson determined that it fitted the standard meter bar 1,553,165 times (and a half). Since then, the standard of length has been redefined in terms of the cadmium wavelengths. In 1907 Michelson received the Nobel Prize in physics for his high precision measurements.

In the ensuing years he produced the largest ruled grating, 117,000 lines in 9.4 inches, and produced the echelon grating by stacking glass plates in a stepped array. In 1919 at the age of 67 Michelson made the headlines with a front page story in the *New York Times*. He achieved the first measurement of the size of a star. In an accurate determination of the angle subtended by Betelgeuse and a guess at the star's distance, Michelson reported the star's size to be 250 sun diameters.

Still driven by the fun of measuring the speed of light, he spent his final years sending beams of light between California mountain tops, relying on the US Coast and Geodetic Survey to determine the distances. At 77 he had another try with an evacuated pipe that provided a 10 mile multiple path length. He was still engaged in his favorite occupation when he died on 9 May 1931.

Michelson never received a college degree since Annapolis didn't grant the B.S. at that time, but he did receive an honorary doctorate from the University of Cambridge and the Rumford and Copley medals from the Royal Society. His Nobel prize in 1907 was granted for accurate measurements, not for the null aether drift. Michelson would probably have felt uneasy about receiving the Prize for the zero aether drift. Professor Charles Townes laid the matter to rest when in 1960 he reported the aether drift as less than one part in 10^{12} using a radiofrequency technique.

What drove Michelson? The work ethic of a struggling father, the discipline of Annapolis, and an intellectual curiosity all contributed. His ability to shift gears with his music and his painting probably kept his personality on an even keel since there is little doubt that his first zero drift determination was a bitter disappointment. A lesser man might have given up.

By the late years of the century light sources had evolved from tallow candles to whale oil to kerosene and to gaslight. All of these proved to yield successively greater light output but only with a greater threat of fire, explosion, or asphyxiation. Thomas Alva Edison took a giant step in better illumination with his electric bulb as well as developing a new scientific technique now called the 'Edisonian approach' - try everything and pray for serendipity.

When Henry Ford's newspaper *The Dearborn Independent* was sued for libel, the prosecuting attorney attempted to discredit Mr. Ford by asking technical questions beyond his ken. Henry said he didn't know the answers but he could quickly hire someone who did.

Thomas Edison could never grasp the intricacies of mathematics, and produced a similar rejoinder about the subject, 'I can always hire a mathematician - they can't hire me.' Henry Ford and Thomas Edison were lifelong friends who did much to bring America's technology to the fore at the start of this century.

Edison succeeded in science through sheer determination - he would try everything! Late in life he sought a substitute for far-east rubber and examined 14,000 native American plants. He concluded that goldenrod contained 5 per cent rubber. Much to the relief of hay fever sufferers the scheme never came to fruition.

His invention of the electric light twenty-two years before Nobel Prizes were awarded has served man as much as any achievement. One may ask whether such a discovery would merit a Nobel Prize today. The accidental discovery of the high temperature oxide superconductor can be no different.

Born in 1847 at a time when America was about to expand towards the California gold fields, Edison came from a poor family but always demonstrated a determination to succeed. As a curious lad he talked a friend into consuming a large number of bicarbonate of soda powders to produce enough gas to make him fly. He talked another victim into eating worms since birds thrived on such a diet. We must certainly marvel at his early powers of persuasion!

At 13 the young Edison worked the overnight train route to Detroit selling newspapers, candy and books. During this long ride he found himself with spare time and brought a collection of chemicals aboard for experimentation. Unfortunately he caused a fire and only barely escaped the loss of his train-riding venture. He did lose his hearing, however, when he received a good boxing around the ears from the conductor, and he remained deaf for the remainder of his life. He found an abandoned printing press which he repaired and installed aboard the train. At each stop he obtained the latest news from the station's telegrapher and would set it in type to sell to passengers.

When young Thomas saved the life of a telegrapher's son by snatching him from the path of a train, he was rewarded with lessons in Morse code and became a roving telegrapher. In his spare time he developed telegraphy as a technique for the automatic recordings of votes. He offered it to the US Senate but was told speed was not important to them. Besides, it might disrupt their filibustering. He lost all his savings on this vote recorder, but he did sell a stock market ticket tape to Western Union for $40,000, and used the money to build a laboratory at Menlo Park, New Jersey. In time he became known as the 'Wizard of Menlo Park'.

His first project was it improve the quality of the Bell telephone, after which he went on to invent the phonograph. The possibility of an electric light next fired his imagination. Home lighting had undergone several phases from 1780 to 1880 but nothing was completely satisfactory. The eighteenth century was dominated by tallow (sheep fat) candles but their poor light output made whale oil lamps more attractive. With the discovery of crude oil in Pennsylvania kerosene lamps replaced whale oil, although the risk of fire was considerable. Gas lighting was next developed, reducing somewhat the risk of fire from spillage but adding the threat of both explosion and asphyxiation. Edison saw the electric light as the solution to all these dangers.

In the Edisonian approach, the 'Wizard' and his assistants tried some 9.000 substances in searching for a viable filament. Platinum was one of the more promising materials but oxidation at elevated temperatures impaired its mechanical strength. He tried encasing it in a vacuum but such technology was crude and the best he could achieve was a very short lifetime. How Edison hit upon carbon fibers is not clear, but he was aware that one could carbonize paper or wood and leave a carbon fiber with a modicum of structural integrity. (This predates our modern technology for making carbon fibers from a polymeric monofilament. The hydrogen can be driven off by heating in an oxygen-free atmosphere and the resultant carbon remains solid to very high temperatures, particularly in a non-oxidizing atmosphere.)

The first carbon filament made from paper and housed in a glass vacuum was lit on 21 October 1879, and lasted 45 hours. Edison arranged to have Menlo Park lit with an array of electric bulbs by Christmas of that year. To cite the press report:

The laboratory of Mr. Edison at Menlo Park was brilliantly illuminated last night with the new electric light, the occasion being a visit of a number of the inventor's personal friends. Forty lamps in all were burning from six o'clock until after ten. The various parts of the system were explained by the inventor at length. As a practical illustration of this method of subdividing the electric current he had two copper wires of about an eighth of an inch in thickness leading to the generating machines placed side by side on cleats along tables nearly the entire length of the laboratory. To these he connected lamp after lamp by merely fastening little wires to each of the parallel supply wires and then attaching them to the lamps. The illumination or extinguishment of any one made not the slightest perceptible difference in the strength of the current. Twenty electric lamps burned with exactly the same brilliancy as did one when the other nineteen were disconnected. The light given was of the brilliancy of the best gas jet, perhaps a trifle more brilliant. The effect of the light on the eyes was much superior to gas in softness, and excited the admiration of all who saw it.

A new feature, shown by the inventor for the first time, was the method of regulating the strength of the current to be used at the central stations. By moving a little wheel the assistant in charge of this branch of the system was enabled to readily vary the strength of the electric lights from the merest glimmer to a dazzling incandescence. When the latter point was reached the little horseshoe paper presented the appearance of a beautiful globe of fire. The method of obtaining the vacuum in the little glass bulbs of the lamps was also explained and proved highly interesting.

Today we use tungsten since it has the highest melting point amongst metals and is reasonably inexpensive. Funnily enough, the advent of the more efficient fluorescent bulb developed by A.H. Compton has not eliminated the popularity of the incandescent bulb. We now use a buffer gas rather than a vacuum, in order to reduce the evaporation rate of the tungsten.

Following the Menlo Park demonstration, the New York Aldermen urged Edison to install electric lights in their city. To respond to this request, Edison had to provide the entire technology including the bulbs, the generators and flywheel control, and even digging the trenches. On 4 September 1882 Edison started to generate electric lighting in New York and for eight years it ran without mishap.

With this success Edison fulfilled the great 'American Dream' and his future machinations were always newsworthy. His goal in life became: 'to do everything to free people from drudgery and create the largest measure of happiness and prosperity.' He moved to larger quarters in West Orange, New Jersey, and began work on the motion picture. The first successful demonstration of movement on film had already been accomplished by Eadweard Muybridge, but Edison took the next step and combined voice and picture. Years later, as the motion picture business mushroomed, unsuccessful efforts were made to break or avoid the Edison patents, partly by employing French cameras and moving to California. His considerable achievement in the development of motion picture technology is best described in the letter he composed in 1925:

> In the year 1887 the idea occurred to me that is was possible to devise an instrument which should do for the eye what the phonograph does for the ear, and that by a combination of the two all motion and sound could be recorded and reproduced simultaneously. This idea, the germ of which came from. . .the work of Muybridge, has now been accomplished so that every change of facial expression can be recorded and reproduced life-size. . . .I believe that in coming years. . .grand opera can be given at the Metropolitan Opera House without any change from the original and with artists and musicians long since dead.
>
> This meant the photographing instantaneously of a scene as viewed by the eye and involved the following problems:
>
> 1. The pictures had to be taken from a single point of view. . . . .In other words, the camera should not move with respect to the background. . . .
>
> 2. The pictures had to be taken at a sufficiently rapid rate to

give a smooth and uniform reproduction without jerking;
3. . . .the interval between successive images would be less
than one seventh of a second. This was a purely physiological
limitation made necessary to take advantage of the phenomenon
of persistence of vision. . . .
4. A carrier of indefinite length was needed and my conception
included taking the photographs on and reproducing the
positive prints from a tape of light, tough, flexible material,
such as a narrow celluloid film.'

Edison goes on to describe his first camera:

My first camera constructed in 1889 and covered by this
patent disclosed the following features which have always
been utilized in the art:
1. A single lens.
2. A long celluloid film carrying a sensitive surface and
having two rows of sprocket holes.
3. A reel from which the film is unwound and a second
reel on which the film is wound after exposure.
4. Mechanism having a minimum inertia for moving the
section of the film between the two reels intermittently
past the lens many times per second, the film being
stopped and brought to rest at each exposure.
5. A shutter coordinated with the feed mechanism to
expose the film during the periods of rest.

Thus the motion picture was born with Edison's camera called the Edison
Vitascope in 1896. But not every idea of the 'Wizard' succeeded - some led to
significant financial losses, such as his magnetic procedure to enrich iron ore.
He did follow up on the discovery of X-rays in 1895 and produced the medical
fluoroscope which he chose to donate to mankind rather than patent it.

When the Great War brought America into the fray in 1917, Edison accepted
Secretary of the Navy Daniels request to consult for them, something he later
regretted for reasons he best expressed himself:

I made about 45 inventions during the war, all perfectly
good ones, and they pigeon-holed every one of them. The
Naval Officer resents any interference by civilians. Those
fellows are a closed corporation. I do not believe there
is more than one creative mind produced at Annapolis
in three years. If Naval Officers are to control the school
the result will be zero. [Edison failed to take note of
Michelson.]

In 1929 Henry Ford erected a permanent tribute to this American genius when he rebuilt an exact replica of the Menlo Park Laboratory at an outdoor museum site in Dearborn, Michigan, called *Greenfield Village*. Edison, by then 82, was invited to the dedication and congratulated Ford on duplicating his old laboratory to 99 per cent accuracy; the one inconsistency, "He never kept it so clean!"

Thomas Edison was the great American tinkerer who felt more comfortable in doing everything himself than in consulting the learned professions, physics or mathematics. He felt intimidated by scientists, although he occasionally found exceptions like Steinmetz, the hunchback wizards of General Electric, who never used equations for this explanations. Few of Edison's employees were scientists, principally competent technicians ever ready to take orders from the boss. If one showed unswerving loyalty to this man and worked hard, one succeeded as part of his workforce. The Edison approach is no longer fashionable, since scientists continually try to reduce their efforts to computer modelling, but in the final analysis dogged determination will always serve one in good stead.

Edison's hearing disability fostered a reluctance on the part of his employees to engage in dialogue with him since this turned into a shouting match. Reporters, on the other hand, found him good copy since a single question would often produce long rambling answers. His deafness prompted the 'Wizard' to repair to the inner workings of his mind where his fertile imagination drove him unceasingly. He died on 18 October 1931. There is one Hollywood film about the man (played by Spencer Tracy) but it doesn't appear to be too popular on the rerun circuit.

Optics took a giant leap in the latter part of the nineteenth century when a soft-spoken, devout Scotsman named James Clerk Maxwell put on his thinking cap (a bonny tam) and elevated the field of theoretical physics to a height only previously attained by Newton. For half a century after his death his personality and impact on optics lay hidden from the public but he has been rescued from oblivion by the recent formation of an International Maxwell Society. His story is unusual:

At the apex of theoretical physics stand names like Newton, Maxwell, Einstein, Schröedinger and Heisenberg, for these scientists have provided revolutionary insight into the workings of nature. Whatever their thought processes, it took a strong determination to discard traditional ideas and to enter uncharted territory. Of these physicists Maxwell's is one of the most difficult personalities to explore.

Not only did he prefer to work alone but his inability to verbalize ideas further isolated him from his scientific colleagues and students. On the other hand, he communicated effectively with himself so that his writings present a clarity and poetry to the reader. This was the man who unified electricity and magnetism into a single theory that helped explain the nature of light, its polarization and its energy - a theory still in use well over a hundred years later.

Some of his inhibitions can be traced to his youth. He was born on 13 June 1831 into an Edinburgh family that claimed lineage from a famous sea captain who swam from a shipwreck by inflating his bagpipes. Arriving ashore in a remote section of India he played Scottish airs to keep the tigers at bay. James Clerk Maxwell cherished the legend and the bagpipes became a family memento.

He spent his youth in the Lowlands, a somewhat mischievous lad who remembered blowing out the candles and tripping the housemaid as she entered the tearoom. With the death of his mother when he was eight, this curiosity-filled youngster was entrusted to an aunt and a cruel tutor. At ten the tutor was dismissed and James was enrolled in the Edinburgh Academy where, like Lewis Carroll, he was so mistreated by his schoolmates that he developed a stutter. He was considered a bookworm, performing odd experiments and brandishing a non-local brogue. His small stature did not help.

While his fellow students acknowledged Maxwell's brilliance, they viewed him with sufficient suspicion to necessitate that his athletic interests be fulfilled in 'loner' activities such as horsemanship, swimming, polevaulting, walking, etc. Left to himself he turned to self-expression in his writing, thinking, poetry and experimentation. His spiritual fulfillment came through his devout religious beliefs. Churchgoing and a daily reading of the Scriptures became *de rigeur* during his lifetime.

Looking back in later years at his school chums he remarked, 'They never understood me, but I understood them.' At 13 he showed poetic ability when he wrote about his school and two members of its staff, Mr. Carmichael and Mr. Williams:

> Ne'er shall the dreadful tawse be heard again,
> The lash resounding, and the cry of pain;
> Carmichael's self will change (O that he would!)
> From the imperative to the wishing mood;
> Ye years roll on, and haste the expected time
> When flogging boys shall be accounted crime.
>
> But come, thy real nature let us see.
> No more the rector but the godess be,
> Come in thy might and shake the deep profound,
> Let the Academy with shouts resound,
> While radiant glory all thy head adorns
> And slippers on thy feet protect thy corns;
> O may I live so long on earth below
> That I may learn the things that thou dost know.

It is interesting to compare this poem with 'My Fairy' by his contemporary Lewis Carroll, both written at the same time and at the same age. Alas, the two appear never to have met since they were at Cambridge and Oxford respectively - 50 miles apart as the crow flies but worlds apart in social interaction.

At this time his father introduced James to the world of learning with visits to the Edinburgh Royal Society. The teenager displayed an interest in geometrical shapes and did some original work. A Professor Fraser even presented Maxwell's work at a meeting of the Society, James being too young to occupy the podium. The work was later published under the title 'Oval Curves'.

Maxwell left the Academy at 16 with a first in mathematics and English. He penned the following lines on this occasion:

> Let pedants seek for scraps of Greek,
>   Their lingo to Macademize;
> Give me the sense, without pretence,
>   That comes o' Scots Academies.
>
> Let scholars all, both grit and small,
>   Of learning mourn the sad demise;
> That's as they think, but we will drink
>   Good luck to Scots Academies.

Still in his teens, Maxwell undertook the problem of the motion of Uranus whose trajectory was perturbed by the yet undiscovered planet Neptune. Colors and prisms intrigued him and he entered the University of Edinburgh to study physics, logic, metaphysics and mathematics.

By 18 his maturity became evident with this remarkable composition:

> I was thinking today of the cognitive faculty. It is universally admitted that duties are voluntary, and that the will governs understanding by giving or withholding Attention. They say that Understanding ought to work by the rules of right reason. These rules are, or ought to be, contained in Logic; but the actual science of Logic is conversant at present only with things either certain, impossible or *entirely* doubtful, none of which (fortunately) we have to reason on. Therefore the true logic of this world is the Calculus of Probabilities, which takes account of the magnitude of the probability (which is, or which ought to be in a reasonable man's mind). This branch of Math., which is generally thought to favour gambling, dicing, and wagering, and therefore highly immoral, is the only 'Mathematics for Practical Men', as we ought to be. Now, as human knowledge comes by the senses in such a way that the existence of things external is only inferred from the harmonious (not similar) testimony of the different senses, Understanding, acting by the laws of right reason, will assign to different truths (or facts, or testimonies, or what shall I call them) different degrees of probability. Now, as the senses give new testimonies continually, and as no man ever detected in them any real inconsistency, it follows that the probability and *credibility* of their testimony is increasing day by day, and the more a man uses them the more he believes them. He believes them. What is believing? When the probability (there is no

better word found) in a man's mind of a certain proposition being true is greater than that of its being false, he believes it with a proportion of faith corresponding to the probability, and this probability may be increased or diminished by new facts. This is faith in general. When a man thinks he has enough evidence for some notion of his he sometimes refuses to listen to any additional evidence *pro* or *con*, saying, 'It is a settled question, probatis probata, it needs no evidence; it is certain.' This is knowledge as distinguished from faith. He says, 'I do not believe; I know.' If any man thinketh that he knoweth, he knoweth yet nothing as he ought to know. This knowledge is a shutting of one's ears to all arguments, and is the same as 'Implicit faith' in one of its meanings. Childlike faith, confounded with it, is not credulity, for children are not credulous, but find out sooner than some think that many men are liars.

He set down his thoughts about the forces in nature, reflecting his early uncertain efforts to categorize them and revealing the long path ahead to his unraveling the relationship between electricity and magnetism:

Of forces acting between two particles of matter there are several kinds.

The first kind is independent of the quality of the particles, and depends solely on their masses and mutual distance. Of this kind is the attraction of gravitation.

The second kind depends on the quality of the particles; of this kind are the attractions of magnetism, electricity and chemical affinity, which are all convertible into one another and affect all bodies.

The third kind acts between particles of the same body, and tends to keep them at a certain distance from one another and in a certain configuration.

His intuition about physics was superb. Remarkably, at 19 he discovered photoelasticity, the process by which transparent materials under stress will rotate the polarization state of light and produce aesthetically interesting pictures that reflect the strains in the material. Maxwell sketched a few such patterns and had them woven into a potholder.

His former classmate Campbell has left his impressions of Maxwell at this time:

James Clerk Maxwell still occasioned some concern to the more conventional amongst his friends by the originality and simplicity of his ways. His replies in ordinary conversation were indirect and enigmatical, often uttered with hesitation and in a monotonous key. While extremely neat in person, he had a rooted objection to the vanities of starch and gloves. He had a pious horror of destroying anything - even a scrap of writing paper. He preferred travelling by the third class in railway journeys, saying he liked a hard seat. When at table he often seemed abstracted from what was going on, being absorbed in observing the effects of refracted light in the finger-glasses, or trying some experiment with his eyes - seeing around a corner, making invisible stereoscopes, and the like. . . .He never tasted wine; and he spoke to gentle and simple in exactly the same tone.

In October 1850 Maxwell left Scotland and entered Cambridge University, the English center for mathematics. He took up residence at Peterhouse, the oldest of the colleges, dating to 1284. Although Cambridge was to stimulate his mathematical prowess, a significant challenge came from the continent with such luminaries as Laplace, Poisson, Cauchy, Fourier, Bessel, Gauss, Euler, Bernouilli etc. What a time to study mathematics! Bringing his various scientific bits and pieces with him, he occupied his spare hours in physics experimentation.

His idiosyncrasies now included running through the corridors, up and down the stairs, and into the courtyard between 2 and 3 AM. The senior dean J.A. Frere admonished him and Maxwell later responded with a poem when the dean left the school for another position.

John Alexander Frere, John,
        When we were first acquent
You lectured us as Freshmen
        In the holy term of Lent;
But now you're getting bald John,
        Your end is drawing near,
And I think we'd better say 'Goodbye,
        John Alexander Frere.'

The lecture room no more, John,
        Shall hear thy drowsy tone,
No more shall men in Chapel
        Bow before thy throne.

But Shillington with meekness,
        The oracle shall hear,
That set St. Mary's all to sleep-
        John Alexander Frere.

Then once before we part, John,
        Let all be clean forgot,

Our scandalous inventions
   (Thy note-lets, prized or not).
For under all conventions,
   The small man lived sincere,
The kernel of the Senior Dean,
John Alexander Frere.'

His writings often displayed an overindulgence in theological philosophy with such bon mots as:

> You may fly to the ends of the world and find no God but
> the Author of Salvation. You may search the Scriptures and
> not find a text to stop you in your explorations. You may
> read all History and be compelled to wonder but not to
> doubt.
> . . .I assert the Right to Trespass on any plot of Holy
> Ground which any man has set apart. . . I think that the
> results which each man arrives at ought not to be regarded
> as having any significance except to the man himself, and to
> him only for a time, and should not receive the stamp of
> society. . . .

He became a member of the exclusive University Select Essay Club, consisting of a dozen students who were referred to as 'The Apostles'. His poetry in language reveals a man with a mission in life:

> Happy is the man who can recognize in the work of today
> a connected portion of the work of life, and an embodiment
> of the work of Eternity. The foundations of his confidence
> are unchangeable, for he has been made a partaker of
> Infinity. He strenuously works out his daily enterprises,
> because the present is given him for a possession.

1855 was an active year for Maxwell. He was called upon to nurse an ailing father; he presented a paper to the Royal Society of Edinburgh on colour perception which extended the earlier work of Thomas Young; he became a Fellow of Trinity College; and he began his most important work by studying Faraday's experiments on lines of force revealed with iron filings and a magnet.

The concept of action at a distance, already utilized by Newton in his study of gravitation, was now being tailored to electricity and magnetism. His metaphysical philosophy emerged time and again when he stated that, 'the only laws of matter are those which our minds must fabricate, and the only laws of mind are fabricated for it by nature,' i.e. the laws of intellect and matter are inseparable. 'The footprints in the sand were his own.'

There is little point in taking the reader through the tortuous path to Maxwell's theory of electromagnetism and his famous equations. It is perhaps sufficient to say that Maxwell devised all sorts of analogies, most of them entirely remote from the situation he wanted to describe, but he justified the procedure in the hope that the mathematical models would ultimately apply. This may be a difficult concept to fathom but we should bear in mind that Maxwell always retained the picture of the lines of force emanating from a magnet or a wire carrying a current, so even if he pictured them as chariots pulling logs through the snow, he never lost sight of his objective. It took Maxwell years to arrive at the correct answer, whereupon he discarded these simple analogies, but his persistence brought success. Since Maxwell occasionally published some of his research along the bumpy road to fruition we are able to follow some of his ideas, but the route has little interest, just as the thousands of materials that didn't work on Edison's path to the electric light. Who cares when success finally arrives? Maxwell's six-year effort produced a final set of equations that were simplicity itself. Like Einstein's special theory of relativity success was self-evident.

After six years at Cambridge Maxwell felt the call of the Lowlands and he applied for a chair at Aberdeen. Shortly after his father's death in 1856, he received the appointment and he left Cambridge. His poetic imagery describes the change from student to professor:

> The transition state from a man into a Don must come at last, and it must be painful, like the gradual outrooting of nerves. When it is done there is no more pain, but occasional reminders from some suckers, taproots, or other remnants of the old nerves, just to show what was there and what might have been.

He expressed his impressions of Aberdeen:

> . . .no one here seems to think me odd or daft. Some did at Cambridge, but here I have escaped. My rule is to avoid the company of young men who I do not respect, unless I have the control of them.
> Gaiety is just beginning here again. Society is pretty steady in this latitude - plenty of diversity, but little of great merit or demerit - honest on the whole but not vulgar. . . .No jokes of any kind are understood here. I have not made one for two months, and if I feel one coming I shall bite my tongue.

During the years 1856-58 Maxwell turned his attention to the rings of Saturn. Limited telescopic resolution could not distinguish structure in the rings and Maxwell tried to solve the problem of the stability of such a structure. He concluded that the rings must consist of small fragments such that any gravitational imbalances would be offset one against the other. A rigid solid ring could easily be drawn into the larger body by any imbalance:

> When we contemplate the Rings from a purely scientific point of view, they become the most remarkable bodies in the heavens, except, perhaps, those still less useful bodies - the spiral nebulae. When we have actually seen the great arch swung over the equator of the planet without any visible connection, we cannot bring our minds to rest. . . .We must either explain its motion on the principles of mechanics, or admit that, in the Saturnian realms, there can be motion regulated by laws which we are unable to explain.

In his *Recollections of Dreamland*, almost Carrollian, he refers to his thought processes:

> There are powers and thoughts within us, that we know not,
> > till they rise
> Through the stream of conscious action from where the Self in
> > secret lies.
> But when Will and Sense are silent, by the thoughts that come
> > and go
> We may trace the rocks and eddies in the hidden depths below.

At 27 he wrote to his aunt to inform her of his impending marriage to Katherine Mary Dewar, daughter of the College principal, seven years his senior, taller than him and somewhat fair. At 34 she was presumably beyond marriage expectations and no doubt her father's role at the college influenced Maxwell. Little else was known of her but she presumably had deep religious convictions of her own since Maxwell's letters stressed this theme.

> If we despise these relations of marriage, of parents and children, of master and servant, everything will go wrong, and there will be confusion as bad as in Lear's case. But if we reverence them, we shall even see beyond their first aspect a spiritual meaning, for God speaks to us more plainly in these bonds of our life than in anything we can understand. So we find a great deal of Divine Truth is spoken of in the Bible with reference to these three relations and others.

He penned a poem of courtship:

Will you come along with me,
    In the fresh spring-tide,
My comforter to be
    Through the world so wide?

And the life we then shall lead
    In the fresh spring-tide,
Will make thee mind indeed,
    Though the world be wide.

About this time the first Atlantic cable was laid but it parted after a few months. On the second try it held for a couple of years. Maxwell produced a few couplets for the occasion, entitled

The Song of The Atlantic Telegraph Company:

Under the sea
No little signals are coming to me
Under the sea,
Something has surely gone wrong
And it's broke, broke, broke;
What is the cause of it does not transpire,
But something has broken the telegraph wire
With a stroke, stroke, stroke,
Or else they've been pulling too strong.

In 1860 two of the colleges at Aberdeen combined, Marischal and King's, and Maxwell's job was eliminated. Apparently marrying the boss's daughter did not prove too influential. He applied to Edinburgh which had just announced a vacant Chair but the position was given to another because of Maxwell's reputation as a poor lecturer. This handicap haunted him throughout his lifetime.

Suddenly without a job, and rebuffed by Edinburgh, he looked with favor at a position that opened at King's College London. One is tempted to suggest that the presence of a Divinity School at the College influenced him. He applied for the Professorship in physics and was immediately accepted in July 1860. However, he developed smallpox before taking up the new post and was nursed back from death's door by Katherine who ministered both medicine and biblical readings. He recovered fully and often expressed gratitude for her efforts. There is little doubt that the Maxwells gained spiritual strength from their religion.

The couple moved into a new house in Kensington, now 16 Palace Gardens Terrace, only a few doors away from a house this author occupied 100 years later. It was a pleasant part of town, near Kensington Gardens, a half hour ride to the College. The park provided adequate opportunity for the riding both Maxwells enjoyed.

Maxwell's work on color, extending that of Thomas Young, was awarded the

Rumford Medal in 1860 by the Royal Society. He converted his attic in Kensington into a laboratory where he continued to work on the subject. He also added an interest in gases when he showed that their viscosity was independent of density. This is only true at low pressures but presumably Maxwell would not have had the apparatus to achieve high pressures. As long as gas molecules are far apart a gas will flow through an orifice with little viscous friction. Only when molecules come very close are the electrons perturbed and influence the viscosity.

Maxwell invented an ingenious method for color photography by employing identical shots through filters of the three primary colours and then projecting the superimposed images through the same filters.

During his London years he completed the final chapter of his work on electromagnetism and summed it up in one sentence:

Light consists of the transverse undulations of the same medium which is the cause of electric and magnetic phenomena.

The medium was, of course, the aether.

With an ostensibly pleasant life in Kensington and with the culmination of his magnum opus on electromagnetism at King's College London, one is slightly puzzled at his leaving after five years, especially with the Royal Society and Royal Institutions nearby. We can only guess that Scotland's call was very strong, although there seems to be some substance to the rumor that Maxwell found the students difficult to control. His lectures were probably well over their heads and he no doubt preferred the quiet contemplation of his own theoretical work. His home at Glenlair could provide this as he was not hard pressed for money, the tenants on the estate yielding income and the local kirk at Parton fulfilling his need for the spiritual atmosphere not readily found in London. In 1865 he resigned from the college and headed north.

At Glenlair he conducted daily prayers for the household and he became an elder at the nearby church in Corsock. Like Lewis Carroll, the pulpit might have fulfilled his career ambitions if it were not for his speech difficulties. Maxwell's physics and religious feelings were intimately interwoven and Divine Guidance operated in a special way for him. One cannot imagine the two subjects to be totally compartmentalized within his psyche.

Two years after returning to Glenlair he and Katherine took a trip to Italy and southern France. By the time of his return he was longing for the academic life again, particularly the close contact with other physicists. He applied for the principalship at St. Andrews but it went to a friend of the Chancellor.

In 1870 his interest in gases was extended to heat and, for his quasi-analog arguments, he introduced his famous Maxwell demon. This was a Carrollian figure the size of an atom that did useful chores like separating molecules into various compartments. It was useful for pedagogical purposes and is still alluded to. Unlike Lewis Carroll, who with Tenniel's drawings has left impressed for all time the delightful images of Alice and the others in Wonderland, Maxwell never described the physical characteristics of his demon. We might expect it to have an ecclesiastical collar, a Scottish brogue, bearded, donned in kilts and tam, and playing those famous bagpipes of Maxwell's ancestor to ward off noisy and raucous students. Maxwell wrote:

> Concerning Demons.
> 1. Who gave them this name? Thomson.
> 2. What were they by nature? Very small but lively beings incapable of doing work but able to open and shut valves which move without friction or inertia.
> 3. What was their chief end? To show that the second law of thermodynamics had only a statistical certainty.

Following the unsuccessful application for the principalship at St. Andrews, Maxwell probably resigned himself to spending the remainder of his life at Glenlair. But in 1870 William Cavendish, the seventh Duke of Devonshire, a wealthy mathematician, donated money to build a laboratory at Cambridge, and Maxwell was approached to take the Directorship and Chair of Experimental Physics. Having completed his *Treatise on Electricity and Magnetism*, he appeared psychologically receptive to the offer. After six years of isolation at Glenlair he revisited the scenes of his student days and as Cavendish Professor began to supervise the construction of the new laboratory. In the interim he had no place to hang his hat:

> Laboratory rising, I hear, but I have no place to erect my chair, but move about like a cuckoo, depositing my notions in the chemical lecture-room 1st term; in the Botanical in Lent (term), and in Comparative Anatomy in Easter.

Maxwell proved an asset to the new laboratory, not only for the attention he paid to the selection of equipment and to building design, but also for the image he brought as a first-rate physicist. It is true his lectures had not improved and his classes were reduced to two or three of the brightest students.

On one occasion at the Cavendish he seemed to suspend his Christian charity when he was asked to referee a paper by Osborne Reynolds, a scientist of first rank who has given us the well-known Reynold's number, a dimensionless quantity useful in aerodynamics and in predicting the onset of turbulence. An excerpt from his report illustrates his lapse:

> Of course I cannot profess to follow with minute attention the course of an acrobat (Reynolds) who drives 24 in hand, but as on more then one occasion he throws up the reins and starts a new team, it is probable that the results will be sufficiently flexible to adapt themselves to the facts, whatever the facts may be....I am afraid I have not answered your letter at all, except about O.R. being the discoverer of dimensional properties in gases. I have always felt inclined to give him leave to practice

at his 'mean range' till he has qualified himself to go in among all comers for the R. (Royal) S. (Society) meetings.

His humor stands up well, as when he complained about a command performance before the Queen to explain the importance of the radiometer and of the famous Otto von Guericke experiment in which a team of horses could not pull apart two hollowed hemispheres from which all air has been withdrawn:

> I was sent to London, to be ready to explain to the Queen why Otto von Guericke devoted himself to the discovery of nothing, and to show her the two hemispheres in which he kept it, and the pictures of the 16 horses who could not separate the hemispheres and how after 200 years W. Crookes has come much nearer to nothing and has sealed it up in a glass globe for public inspection. Her Majesty however let us off very easily and did not make much ado about nothing. . .

The Crookes radiometer is a well-known device consisting of foils mounted on a rod, silvered on one side and blackened on the other, and free to rotate in an evacuated tube under the action of sunlight. For a time it was believed to be an example of radiation pressure, a subject to which Maxwell made a contribution. His comment on Alexander Bell is a mocking tribute to the phone:

> One great beauty of Professor Bell's invention is that the instruments at the two ends of the line are perfectly alike. . . the perfect symmetry of the whole apparatus - the wire in the middle, the two telephones at the ends of the wire, and the two gossips at the ends of the telephone, may be fascinating to a mere mathematician, but it would not satisfy the evolutionist of the Spenserian type, who would consider anything with both ends alike. . .to be an organism of a very low type. . . .

Maxwell's response to a paper about soap bubbles underscores his potential as an author:

> On an Etruscan vase in the Louvre figures of children are seen blowing bubbles. These children probably enjoyed their occupation just as modern children do. Our admiration of the beautiful and delicate forms, growing and developing themselves, the feeling that it is *our* breath which is turning dirty soap-suds into spheres of splendour, the fear lest by irreverent touch we may cause the gorgeous vision to vanish with a sputter of soapy water in our eyes, our wistful gaze as we watch the perfect bubble when it sails away from the pipe's mouth to join, somewhere in the sky,

all the other beautiful things that have vanished before it, assure us that, whatever our nominal age may be we are of the same family as those Etruscan children. . . .

In 1877 he showed the first signs of abdominal cancer and died two years later. He is buried in the cemetery at Parton in Scotland. A recent memorial was a gift of the International Maxwell Society.

One of his biographers, Ivan Tolstoy, attempted to explain why Maxwell faded from view for almost 100 years. He was never knighted and little has been done to rectify that oversight. In particular, it is difficult to understand why the Queen's speech in 1960 on the tercentenary of the Royal Society failed to mention Maxwell amongst its distinguished members. Being a loner has its pitfalls.

If physics hadn't satisfied his creative drive it is probable that his writing would have achieved great heights. Lewis Carroll, a modest mathematician, was catapulted to fame by *Alice in Wonderland*. A chance digression of Maxwell into the literary field might have met with similar success. His Achilles Heel lay in verbalizing his complex thinking, although his faithful dog Toby always sat nearby and appreciated his musings. Undoubtedly he and Katherine communicated through the Scriptures. Richard Feynman is reputed to have called Maxwell's contribution to physics the greatest of the nineteenth century. I would not care to disagree with such an authority.

During Maxwell's lifetime a far-reaching revolution in aesthetics reared its head to the resentful screams of the public and the art critics when Impressionism was born. One of the great masters of the technique was Claude Monet.

In June 1868, after being driven from an inn in Fecamp, France, Monet tried to end his life. Such is the driving force of creativity that artists can find life impossible if deprived of their ability for creative output. Starvation has also driven many to the grave.

Monet was so obsessed with his craft that he would paint the same scene as many as 50 times, striving to capture it in different lighting. Imbued with the play of light and color from the sun, Monet spent his entire life painting out of doors, a departure from accepted practice. It required coping with the elements and having to face an everchanging reflectivity. Monet sometimes waited an entire day for the right angle of sunlight.

While Rembrandt seemed to be driven by the psychological effects of light and dark, Monet was as passionately stirred by the influence of color. Much has been written about the Impressionists and the clever use of complementary colors that blend within the human optical system. The same technical blend could be achieved by thoroughly mixing the colors before applying them, but it is not the same. The contrasting colored segments must be resolved by the eye- the aesthetics are beyond rationalization.

Monet and Rembrandt 'cheated' by producing paintings that distorted the physical facts. They learned their prestidigitation after years of empiricism - it did not rely on the magician's diversion or speed of movement. Rather, the viewer must concentrate his efforts on a stationary theme. The entire world of abstract art has followed this mystique of distortion.

Monet was born in 1840 and grew up in Le Havre. The sea filled his early views of the world. In his own words,

> I was born undisciplined. Never could I be made to obey a set rule. School was always a prison to me. I could never bring myself to stay there while the sun was shining and the sea was tempting. It was such fun scrambling over cliffs and paddling in the shallows. To the despair of my parents, such was the unruly but healthy life I lived until I was fifteen. At school I doodled in the margins of my books, and I decorated our blue copy paper with ultra-fantastic drawings. I drew the faces of my schoolmasters as outrageously as I could, distorting them out of all recognition.

At 15 Monet sold caricatures that were displayed in a framemaker's shop. The artist Boudin 'discovered' Monet through these drawings and introduced him to landscapes. Although he was twice Monet's age, the two colleagues painted outdoor scenes together. Having saved a little money from his caricaturing, Monet was lured to Paris but shortly afterwards he was conscripted for two years' military service in Africa.

Fascinated with the scenery in Algeria, he began some serious sketching in his spare time. However the climate didn't agree with him and he returned to France to continue with his landscapes and seascapes. 'I'm so eager to paint everything I see, that it's driving me crazy! My head's bursting with it all.' The problem raised by his desire to paint ever larger canvases was solved by digging a trench in the ground to lower the canvas and enable him to reach the upper sections.

The gypsy life in moving from scene to scene forced him to rely on his meager income from the sale of his work. Unfortunately, his style was too radical and sales too infrequent. Often he was forced to leave canvases with landlords in lieu of rent. To avoid having his works seized and auctioned off to pay his debts, he preferred to destroy them. Although his parents were well off, they considered him a waster and refused to come to his rescue. One of his girl friends, Camille, bore him a son, but he was unable to provide for the two. At this time he reached a low point of despair and attempted suicide.

He was rescued by a Monsieur Gaudibert who commissioned him to do a portrait of his wife, and Monet was able to settle down for a while, even joining Renoir in painting landscapes. With the outbreak of the Franco-Prussian War he escaped to England but his work met with little acceptance there. He did a number of paintings of the Houses of Parliament, but were it not for a refugee art dealer he knew in London, Monet claims, 'I would have starved to death.'

After his return to France, the Impressionists Pissaro, Cezanne, Renoir, Sisley, Degas and Monet organized a group show but found their technique was ahead of its time. In 1872 a painting of Monet's entitled 'Impression; Sunrise' (on back cover) was panned by a critic and led to the permanent label 'Impressionism' to describe the group's technique. Through dogged determination the group slowly gained acceptance.

Monet lived to 86 and to see the day his painting 'The Water Lillies' would occupy eight wall panels of two oval-shaped rooms in the Musee de l'Orangerie in Paris. It was inspired by his own garden of lillies. No trip to Paris is complete without seeing it.

Other well-known paintings include the facade of the cathedral at Rouen painted at various times of the year, myriads of haystacks, poplar trees and snow-covered landscapes. When he painted the St. Lazare railway station he was more interested in the effect the steam had on fogging up the view than in the locomotives themselves.

Abstract painters have carried the revolution further. They no longer attempt to record any recognizable image. In 1913 the Cubist Braque cautioned the public, 'How can you see a *nude descending a staircase* when there is no nude and no staircase?' The appeal of abstract painting is through a direct transfer of feeling through color and shape, without the intellect consciously intervening. How this psycho-visual process works is a mystery, but its importance dominates our lives. No day passes without each of us viewing a deliberately distorted man-made image.

One of man's important discoveries at this time proved to be high energy gamma rays (100,000 ev or so). In 1896, Henri Becquerel discovered radioactivity - his observation of the rays from uranium casting a shadow of a key recorded on a photographic plate is a famous image taught in schools. Röentgen, a year earlier, produced high energy radiation by bombarding metals with high speed electrons. The device employed by Röentgen could be made portable and provided copious sources, called X-rays, for an internal inspection of bones, etc. X-rays are still an important diagnostic technique, although they are gradually being replaced with less destructive techniques such as magnetic resonance, optical fibers, and ultrasound. At first it was not realized that the benefits of X-rays could be overshadowed by their cancer-causing destructive capability.

In passing through matter and people X-rays and gamma rays displace electrons and destroy cells, particularly insidious in living tissue. The body possesses an immune mechanism that recognizes broken cells as 'trash' and eliminates them through the lymph nodes Alas, if too many cells are destroyed the body may not recover since nuclei for carcinogenic cells are also created.

The reverse process is also possible i.e. cancerous cells can be destroyed by X-rays at a higher rate than normal cells. This therapeutic approach to cancer treatment is still employed today although other avenues like chemotherapy are becoming increasingly popular.

A virtually unheralded achievement in the 1890's was the brilliant work of G. Lippmann in France who demonstrated how to make color photographs with black and white emulsions only (no colored filters or dyes were employed). By backing the emulsion with a metal mirror the reflected light was 180 degrees out of phase with the incoming light (such is the property of a metal) and an interference pattern developed that depended on the color (wavelength) of the light. The silver salts in the emulsion were darkened where the two waves interfered constructively, leaving a 3 dimensional, wavelength-dependent diffraction pattern recorded in the film. After removing the metal mirror and developing the film, the image could be projected on a screen with a white light source and the various colors would be vividly reconstructed.

Even though Lippmann won the Nobel Prize for physics in 1908 the technique has never been exploited commercially. Why? It became easier to employ colored dyes in emulsions.

By the end of the nineteenth century man's understanding of light had left enough problems unanswered to arouse interest amongst all scientists, even though research funding was limited to a few large companies and government labs. What did we believe about light at the turn of the century?

Light was propagated by an aether, even though Michelson failed to find the aether drift. Both the wave and corpuscular theories place scientists in two camps - like Republican and Democrat. The wave theory successfully predicted diffraction patterns when light passed through slits of comparable size to the wavelength of the light. Polarization suggested two states of light, capable of separation by a doubly refracting material like Iceland spar. The velocity of light had been measured by Michelson in air and in water by employing the Foucault rotating mirror technique. The spectrum of sunlight both in the visible and in the heat-producing infrared had already been measured by William Herschel; glass lenses free from astigmatism had been produced by combining flint and crown glass, their indices of refraction differing. The excitation of gases produced discrete optical lines as did the absorption. Diffraction gratings were accurately ruled and permitted light to be more effectively dispersed into its various wavelengths, advancing the field of spectroscopy and providing a correct determination of the wavelength of specific lines emitted by gases. Maxwell's equations had unified the theory of light into its electric and magnetic constituents. It was recognized that the blueness of the sky arose from the less effective scattering of the longer wavelength reds and yellows from the sun, already postulated by Leonardo. The motion picture camera had been developed by Muybridge and Edison, both employing the emulsion recording technique of Fox Talbot. The electric light, the recognition of the eye's focal anomalies and the means of correction with eyeglasses, and the first experiments in the use of long wavelength optics (radio) to convey messages by wireless brought man into the twentieth century possessed of a firm feeling of scientific triumph.

Yet, even as the twentieth century draws to a close, we are reminded that nature is always more complex that we imagine and it is but a fool who would suggest that we ever totally understand the vagaries of nature. Our intellectual curiosity has led to technological achievements that appear to improve man's lot - unfortunately we haven't a clue as to where we're heading. At the end of the twenty-first century we shall be further reminded 'what fools we mortals be.' The laser and semiconductor technology could not have been anticipated on January 1, 1900, although wars and pestilence could. Certainly the monopolization of man's time by the cinema and TV could scarce be predicted. The exploration of space and the ability to examine the outer reaches of the universe grew at a rate far beyond the imagination of authors like H.G. Wells. The history of science has demonstrated that the Creator is far more subtle than we can conceive, and there is no reason to believe that this philosophical argument will be altered.

CHAPTER IV

Early 20th Century
Part 1

Let us raise the curtain on the twentieth century by describing the antics and optical research of Robert Williams Wood, a physicist and optical engineer par excellence and a showman of the ilk of P.T. Barnum.

In the one year 1903, eight years after Röentgen's first report on X-rays, 100 papers appeared on N-rays, a discovery of Professor René Blondlot at the University of Nancy in France. A fascinating feature of this new phenomenon was the claim that the human brain emitted these rays. While the emanations were weak their existence would assure Blondlot eternal fame. The French Academy of Sciences awarded him a gold medal and 20,000 francs, and his name became newsworthy. Mankind, the press claimed, was on the crest of a new wave, potentially greater than the contribution from X-rays.

Professor Wood, noted optics expert from Johns Hopkins University in Baltimore, was unable to confirm the N-ray discovery, and on a visit to Cambridge, England, Heinrich Rubens encouraged him to visit France and take a close look at Blondlot's work. No English, German or American scientist was able to duplicate these results, virtually all published papers originated from French laboratories.

The noted American professor was welcomed by Blondlot to a demonstration of these mysterious rays, which could penetrate everything except wood (no pun intended). Ushered into Blondlot's laboratory, the lights were extinguished for dark adaptation and Wood was directed to gaze through the eyepiece of the apparatus. Wood could see nothing. Handing the American a piece of wood to interrupt the beam, Blondlot would call out when this happened. As Wood expected, the Frenchman was shooting in the dark. When Blondlot's assistant took over this chore and both Frenchmen were so engaged, Wood removed a key prism from the apparatus and amusedly watched as the observations continued. Before the lights were turned on, Wood restored the prism.

Back in England, Professor Wood sent a letter to Nature describing the Blondlot observations even in the absence of the key dispersing prism or the wooden shutter held by Wood. N-rays were buried by Wood's guile and Blondlot was discredited, dying in 1930.

Professor Robert Wood was born in Concord, Massachusetts, in 1868 and during his lifetime never behaved other than like a mischievous boy. As practical joker, debunker and genius in optics he developed a world-wide reputation. His lectures were masterpieces of showmanship and his optical experiments made important inroads during the early days of quantum mechanics. He was referred to as the 'physicist who never grew up.'

As a tot being reared in Concord, he was bounced on Ralph Waldo Emerson's knee. When the family moved to Jamaica Plain, a southern suburb of Boston, his talent for science was encouraged by a neighbor who operated a blower factory and permitted young Rob free reign of the tools and chemicals.

On one Memorial Day the young teenager placed a dunce cap on the head of a 15-foot high statue of a Civil War Soldier and the celebration parade was held up while the fire department brought in the 'hook and ladder' to remove the cap. Robby discovered that a mixture of chlorate of potash and sulfur made a loud explosive. Fabricating a weighted pile driver actuated by a rope, Robby scared horses, shattered windows, and frightened people. He discovered an explosive sensitive to touch and sprinkled some on the back fence to keep cats off.

He made the headlines by fabricating a fossil of a huge bug and 'planting' it in a local abandoned quarry used by Harvard anthropology students for their digs. Forced to walk through a tough neighborhood, he took some sodium along. Passing in full view of a gang of hostile lads, he spat into a puddle and at the same time used sleight-of-hand to drop the sodium on the same spot. The smoke and fire earned him the sobriquet Satan. Robby was fond of slingshots, fireworks, telescopes, and microscopes in that order.

The Chicago Tribune was taken in when it published a letter of Wood's about a shining piece of blue metal that struck the earth like a fireball. Included was the fabricated chemist's report confirming the metal to be a new element. He was only 19 when he achieved this national attention, albeit anonymously.

Wood published his first paper shortly after he entered Harvard in 1887, when he reported on an experiment he performed that disproved a theory of glacial erosion propounded by a geology professor at the school. After graduating in 1891 he moved to Johns Hopkins University where he took up residence in a boarding house. An argument with his landlady as to whether she was using leftover steak for the hash that was sure to follow the next day was settled when Wood sprinkled lithium chloride on his unfinished meat and performed a flame test the next day to identify the lithium in the hash.

He fabricated a nine-foot long megaphone that enabled him to focus his voice from the boarding house roof across the road on to the ears of passersby. Such comments as, 'Pardon, you've dropped your garters' would evoke howls from his friends and bewilderment from the pedestrians.

He accompanied his future wife on a winter sleigh ride, impressing her with a hot water bottle filled with an exothermic mixture of sulphuric acid and icewater. On their honeymoon to Alaska in 1892 they stopped at Yellowstone National Park and Wood covertly added fluorescein to the famous Emerald Pool, shocking the guide who had never witnessed such an inexplicable and rapid change of color.

Wood entered the University of Chicago as an Honorary Fellow in 1892 and performed duties as a lecturer's assistant. Employing a large oilcloth with a hole, he devised a technique to blow massive smoke rings several feet in diameter that could bowl over cardboard boxes and be fired at the audience. In 1894 the Woods journeyed to Berlin where his interests began to shift from chemistry to physics. He was present in Germany when Röentgen discovered X-rays and he provided the American newspapers with their first reports.

A typical example of Wood's swashbuckling style occurred while under arrest for taking a short cut through a railroad tunnel during an Alpine climb. He escaped the police by sliding down a hill. In 1896 Wood learned how to fly the first hang glider.

In another prank he boarded the second-class compartment of an 'el' train in Berlin with a proper ticket in his pocket but waving a third class ticket for all to see. Accosted by a guard, Wood pretended he could not understand German and engaged in a 20 minute verbal battle to the delight of the passengers. Arrested at the next stop, he extracted the correct ticket to the amusement of all, and waved a parting *Auf Wiedersehen* to the perplexed guard.

Pretending to be a reporter for the *San Francisco Chronicle*, he rode the new Trans-Siberian railway free in 1896. A friend of Tolstoy inveigled Wood to smuggle a few copies of a German edition of the author's banned works into Russia. Strapping the books to his body, he foiled the customs inspector and delivered the copies personally to Anton Chekhov, a friend of Tolstoy. Wood's ignorance of Russian forced him to sketch whatever he wanted to buy in a shop. The first merchant he approached thought Wood was selling sketches and turned him away.

Back in America Wood constructed the first underwater bathysphere for subsurface marine viewing, employing a surface-operated bicycle pump for air. In 1897 he joined the faculty at the University of Wisconsin and constructed an even larger smoke ring generator, this time employing invisible rings of air to knock things over. The everlasting gratitude of the university came with his invention of an electric heating unit to thaw out frozen water pipes during the severe winter of 1899. The state legislature was so grateful it appropriated $200,000 for a new university building.

Wood's work on the emission spectrum of sodium vapor and his improved version of the Rowland diffraction grating set the groundwork for Bohr's theory of quantized orbits. With Rowland's death in 1903, Wood was invited to take the vacant Chair at Johns Hopkins University and he remained in Baltimore until his death.

The following year he vacationed on the Left Bank in Paris, joining the famous artists in discussions of technique. Accepting Lord Rayleigh's invitation to speak at Cambridge and spend the weekend as his guest, he brought along his lecture apparatus of filters, lenses, and prisms wrapped in men's torn shorts and soiled shirts. While paying his respects to the host in the lounge, the butler unwrapped the lecture items, stored the torn clothing in the dresser, and lined up the optical equipment with Wood's hair brush and razor on top of the dressing table!

Wood submitted the manuscript of his popular book *Physical Optics* to his publisher in 1905 and was in Paris when the three sets of proofs arrived. He corrected and sent off one copy, discarding the others in the trash. On his motor journey around France he discovered one hotel with sheets of his proofs hung in the 'loo' as toilet paper.

He bought a summer place in Easthampton, Long Island, and received some press coverage for his rotating mercury mirror. The mercury, sitting in a dish at the bottom of a well, formed a parabola during slow rotation and served as a reflecting telescope. A Texas group offered the famous professor $50,000 to build a huge rotating mirror in the Lone Star State to communicate with people on Mars. Wood advised them that a large black spot on the Bonneville salt flats would be cheaper and just as visible. Fame was producing unwanted attention.

Wood suggested projecting a slide of a painting on to the painting itself to enhance the light and dark contrasts, a trick picked up by art galleries. In 1909 he received the Rumford medal.

At an outdoor party at his Easthampton home Wood dressed as Bleriot, the French pilot who had flown across the English Channel. Running an invisible line from the roof of the barn to a distant point, Wood mounted the roof and indicated he would fly a small plane he had made. Disappearing from sight momentarily, Wood sent a flaming plane along the wire to a fiery crash. The women were scared out of their wits and Wood took his curtain call on the rooftop.

He constructed a 42 foot long tube, seven inches square, at his summer place, with a six inch lens that had a 42 foot length. When he returned the subsequent summer to complete the spectroscope, the tube was full of cobwebs. Grabbing the cat, he shoved it in one end and closed the cover. Upon emerging at the other end the cat had performed a thorough cleaning job. The diffraction grating had 75,000 lines, 15,000 to the inch and the entire visible spectrum was 50 feet long! Dissatisfaction with not being able to keep out moisture forced Wood to construct an underground tube. The mason he engaged assumed he was building a cesspool, and couldn't understand why it had to be so straight.

In 1913 Wood received an honorary LLD from Birmingham University, sharing the honor with Lorentz and Madame Curie. He wrote an early Sci-Fi book entitled *The Man Who Rocked The Earth* and made some models of moonlike sequences that he photographed to illustrate Jules Verne's books. At the famous 1913 Solvay Congress, limited to the most distinguished names in physics, an advance request from Madame Curie to refrain from smoking was honored up to the point when Wood lit his pipe. Having been provided with fine Havana cigars by Solvay, the conferees felt cheated and soon all were smoking. Madame Curie, the only woman, stormed out of the hall.

At the start of the Great War both Wood and Edison were brought in as Navy consultants. Wood advised the High Command to employ blisters on ships to defeat torpedoes and also suggested trained seals to locate submarines. The Admirals called this latter idea 'more screwy than the tail end of a ship,' but the British tried it out by rewarding starved seals when they located a target. In actual practice, though, the hungry seals wandered off at the sight of fish. Commissioned a Major and sent to France, Wood worked out the course of the transport ship by employing a home-made sextant, much to the chagrin of the captain who considered the course secret. To foil Wood, the Captain set the clocks ahead 45 minutes. The Professor later met General Pershing who asked him his mission. Wood replied, 'Actually, General, it's secret but I suppose there's no harm telling you.'

In France, Wood developed a signalling telescope that could be seen a mile away but focused to an area of only a few square feet. He also demonstrated to the authorities a method to detect false passports with ultraviolet light.

After the war Wood moved his 40-foot spectroscope to a new laboratory in Schenectady and studied the magnetic splitting of spectral lines. His international reputation grew and on a visit to King Tut's tomb in 1931 he was covertly handed some pieces of purple jewelry from the tomb. The mystery centered on how the Egyptians produced this rich color with gold. Back in Baltimore Wood pinpointed the color as due to an iron-gold alloy. He was even called in to solve murders and to debunk psychics.

Having never completed his thesis work at Chicago, he was not entitled to be called a Doctor. In 1931 the University of Berlin decided to award him an honorary degree, but only after the board were shown Wood's book of sketches *How To Tell The Birds From The Flowers*.

In 1935 Wood was elected president of the American Physical Society and in 1938 he was awarded the English Rumford medal at which ceremony he broke the non-smoking rule in the Royal Society. After that, the rule was relaxed.

At 70 he became enamored of the boomerang and at a Johns Hopkins football game entertained the stands at half time. One spectator held up an umbrella only to have it snatched from his hands by the boomerang. With his death in the fifties physics lost its clown prince. His contributions to optics later proved important in helping Bohr, Goudsmit, Schrödinger, Heisenberg and Pauli compare quantum theory with experiment.

Before continuing our travels through the present century we shall change course by telescoping what the twentieth century has taught us about the two fundamental particles, photons and electrons. The interaction of these two makes the world what it is. We may treat them as separate entities because they appear to be so but they merely represent two ways that nature packages energy, the photon at the lowest end of the scale, the electron at the next level. Still higher are atomic nuclei containing protons and neutrons but it is not essential to place any emphasis on these two if we are primarily painting a picture about light. Except for the Mössbauer Effect photons and nuclei barely interact.

CHAPTER V

Tutorial 1
The Photon and the Electron

A - Structure:

Man has evolved bathed in sunshine for half his life. His eyes, sensitive to the colors emitted by the sun, utilize light to guide his movements, recognize his friends and enemies, permit him to engage in sports, and, when the sun goes down, enable him to watch a TV set that creates its own colors. The sun also emits infrared light to keep him warm and ultraviolet light to give him a sun tan. Over the past thousands of millennia man emerged from some primitive form in a liquid-like environment to grow into the sophisticated creature we now recognize, capable of writing complicated treatises on his philosophy, his aspirations, and the nature of light; and destined to populate the universe far beyond his own planet. He is probably the only creature in our galaxy interested in buying a copy of this book.

In both a philosophical and scientific sense we don't know what light (or a photon) is since we have no way of examining its pieces, but we do know something about its behavior. For one thing, it never stays put, always moving at the phenomenal speed close to 186,000 miles a second, no more - no less. When moving through glass it appears to slow down but this is because it is continually being absorbed and re-emitted by the electrons, each such event creating a time delay. When it travels in any given direction in free space, its electric and magnetic sensors are rapidly oscillating perpendicular to this direction, the one growing while the other is waning.

But we seem doomed not to be able to 'see' photons in the way we see most things i.e. by shining light on them and observing the reflected rays. For one thing, photons rarely bounce off each other - they can be sent in opposite directions along an optical fiber without taking any notice of each other. When the eye sees photons, they must be destroyed by absorption in our retina to provide the appropriate information of color, position, and time of arrival, i.e. we have to shoot the messenger to learn what's happening. The color (wavelength) and energy depend on the frequency of vibration of the electric sensors. The only other property of commercial importance is the direction along which the electric sensor vibrates, any of the 360 degree intervals around the circle whose plane is perpendicular to its direction of propagation. This is called the polarization direction.

Photons can be considered one of the Creator's most popular methods for packaging energy. Man has utilized photons over a very large range of their energies, from low-energy radio waves to high-energy gamma rays, the latter a million billion times more energetic than the former. I use the word energy freely because everyone understands it in terms of his own experience, and why he must periodically drive to a gas station or to McDonalds to refuel. At present we don't fill the auto tank with photons because gasoline is more practical, although we have seen pictures of cars running on sunshine that utilize silicon absorbers to convert the sun's rays into electricity. We also know that plants absorb sunlight for energy and we in turn use this energy in the food we consume.

Physicists are loath to venture a guess as to why photons were created in the scheme of things - this restless will-o'-the-wisp entity moving at the maximum velocity known and oblivious to his fellow photons. But man and nature make ample use of them, from radio waves for conveying information to X rays and gamma rays for "seeing through things."

B - Energy, Momentum, Frequency, Polarization:
Our picture of the photon can be clarified by recognizing the simple relationship between the frequency of the electric and magnetic oscillations and its energy. The are, in fact, identical when we multiply the former by Planck's constant to convert into the appropriate units. For example, photons that we recognize as red light have a frequency in the range of 400,000,000,000,000 hertz (vibrations per second) or an energy of 1.6 electron volts (eV.) It's difficult to conceive of anything oscillating so rapidly, but red photons do it all the time.

Suppose we slow the oscillations way down to where we are on more familiar ground, like the radio waves whose frequency is marked on the dial. My daughter's radio is tuned to 1000 kilohertz i.e. one million oscillations per second. It picks up these waves from the air, recognizes the information stored into groups of them, and blasts out rock and roll with a power assist from the electric company. But even a million oscillations a second is difficult to fathom. If there are photons in the universe whose frequency is slowed down to a rate we can more readily envision, like one hertz, they would be impossible to detect.

The 1.6 eV energy of the red photon may not mean anything in terms of your daily experience but if you had a cup of water and arranged that every molecule suddenly absorbed a red photon, the water would be instantly raised to a temperature of 10,000 degrees, exploding violently, taking the roof with it, and silencing the radio. (Leonardo designed a gun which was loaded, heated red hot to about 600 degrees at the breech end, and then filled with water. The rapid conversion to steam drove the ball about a mile.) On the other hand, if each water molecule were to absorb a one kilohertz radio wave, it would only heat the liquid a fraction of a degree.

The photon also has momentum, a concept familiar to billiard players. You determine its momentum by dividing the energy by the speed (velocity of light). We don't ordinarily concern ourselves with photon momentum because it is so small but if each surface atom of a perfectly reflecting billiard ball were to be simultaneously struck by a red photon from a powerful laser, the ball might achieve a speed of a foot per second. That would be significant for a carom shot but it would take an unbelievably intense laser to achieve this (a laser version of billiards is unlikely to be available in the near future). A similar collision with radio waves would scarcely be observed even under a microscope. (In spite of its small value, the momentum of photons was confirmed in 1923 by the Nobel Laureate A.H. Compton who scattered 17,000 eV X rays from individual electrons, playing a sub-microscopic game of billiards between photons and electrons).

A property closely associated with the frequency of oscillation is the photon's wavelength, i.e. the distance it travels in making one oscillation - simply divide the speed of light by the frequency. For the red photon this is about 700 nanometers, or slightly less than one millionth of a meter (split a human hair into a hundred strands). A one kilohertz radio wave has a wavelength of about a half kilometer, the microwaves in your oven are about a centimeter in length, and an X ray is about the size of an atom.

Another property that we find useful in yielding a mental image is the photon's polarization. There are techniques for arranging that all photons travelling in one directions have their electric oscillations in the same direction. For example, if a beam of light were travelling toward you the electric oscillations would be in the plane of this page and all pointing toward the top of the page. When this occurs the light is said to be linearly polarized, a property utilized in polaroid sunglasses which absorb photons with the wrong polarization.

C - Particle or Wave?

From the time of Newton until the early part of the twentieth century the controversy reigned as to whether the photon was a particle or a wave, you were either in one camp or the other. The bending of light as it goes through a lens suggests a wave while the photoelectric effect (photon strikes a metal and knocks an electron out) appears more amenable to a particle concept. As it turned out both were logical descriptions, the photon being a two-headed animal that *appears* to behave differently depending on the situation it encounters (people are no different). As photons are put through increasingly exotic technological paces, as in lasers and optical fibers, we shall, no doubt, discover them to be even more versatile and be likely to endow them with further properties. For the present we say that they are interchangeably particle and wave.

If the photon behaves like a particle does it have mass? In order to answer this we would have to weight it, i.e. determine how strongly it is attracted by the pull of gravity. But how do you grab something travelling with the speed of light and weight it? This measurement turned out to be one of the milestones in physics when the astronomer Eddington followed a suggestion of Einstein's and detected the gravitational deviation from a straight-line flight path as the light from a star passed close to the sun on its way to earth. This was ingeniously measured in 1919 by blocking out the light from the sun with the moon - otherwise known as a solar eclipse. The deviation was only a few seconds of arc but it brought

Einstein world fame.* (cf. front cover).

More recently an accurate measurement of the increased energy of a photon as it fell in the earth's gravitational field was made by R.V. Pound at Harvard.

Since the momentum of a photon is its energy divided by its velocity and the momentum of a billiard ball is its mass times its velocity, by analogy we can divide the photon momentum by its velocity to determine the 'mass'. If we do this bit of arithmetic we find that a red photon only weights one billionth as much as the lightest atom (hydrogen) and a radio wave is too light (pun intended) to take seriously. The photon represents the Creator's scheme for packaging energy with very little mass and momentum. Like a small sports car - it's very light and built for speed.

D - The Velocity of Light

The speed of light is vital in the make-up of things. We know that we can't accelerate electrons, atoms, or spaceships up to this velocity, at least not as yet. It would require wild speculation to propose what is on the other side of this speed barrier, if attainable. It is an interesting result of Einstein's relativity that two spaceships travelling at $3/4$ of the velocity of light and approaching each other in opposite directions do not pass at one and one half times the speed of light. If you are a traffic cop in one of the ships and use a radar beam to measure the speed of the other, you could not arrest them for exceeding the universal speed limit of the velocity of light. You would notice, however, that the return radar signal was shifted to higher frequencies, having picked up energy bouncing off the moving spaceship. Since photons do not increase energy by going faster they do so by increasing their frequency.

This photon manifestation of the well-known Doppler shift that changes the frequency of the whistle of a passing train has provided us with the astronomical observation of the Hubble red shift. Light from distant stars is shifted to lower frequency, suggesting an expanding universe. Extrapolating back in time; if this is a faultless scientific maneuver, leads to the conclusion that the universe started with a Big Bang.

It's not difficult to envision the Doppler effect. Imagine yourself able to count the frequency of the electric vibrations of a photon as it passes you. If you then get in your car and drive toward the photon, the frequency of oscillations passing you will increase, and versa versa if you move away.

There are other subtle phenomena associated with the velocity of light such as the twin paradox. If one of a set of twins embarks on a space voyage approaching the velocity of light he will return younger than the other. Basically, time, space, and mass all change dimensions as one approaches the speed of light. It is not an easy subject to comprehend, even scientists dispute their conclusions, hence I recommend you drive slowly lest you shorten your life span trying to lengthen it.

---

* Shortly after this experiment was given world-wide publicity, a cartoon appeared in an English publication showing a bobby catching a thief by bending the light from his flashlamp around a corner. Today we can easily do this with optical fibers.

E - Converting Photons into Electrons

In addition to the velocity of light, there is another critical barrier at which photons are regulated. When their energy exceeds a million electron volts they prefer to convert this energy into mass through a process called pair production, the photon disappears and an electron and a positron appear. It does this in the vicinity of an atom in order to take up the "kick". (This has to do with conservation of momentum.) The energy of the electron and positron are each one half million electron volts, as given by Einstein's famous equation $E = mc^2$. Hence we come to the second stage in the Creator's method for packaging energy - the electron and its rarely seen sister, the positron.

F - Summary

PHOTON PROPERTIES

1. Energy = Frequency of oscillations times Planck's constant.
2. Momentum = Energy divided by Velocity of Light.
3. Mass = Momentum divided by Velocity of Light.
4. Wavelength = Velocity of Light divided by Frequency.
5. Electric and Magnetic Oscillations are at right angle to each other and perpendicular to its direction of motion.
6. Velocity = 2.9979×10^{10} cm/sec; 186,281 miles/sec.

NOMENCLATURE

1. Gamma rays - the highest energy photons, from about half a million eV upwards.
2. X rays - photons in the energy range from about 100 eV to several hundred thousand eV.
3. Ultraviolet - From about a few eV to 100 eV.
4. Visible Light - Around 1 eV.
5. Infrared, Microwaves, Radar, and Radio Waves are, in that declining order, the photons at the lowest energies of the spectrum.

THE ELECTRON

A - Structure

The electron is an entirely different animal than the photon. It can sit still, although here on earth it spends most of its time moving around atoms and fulfilling the important role of holding us together. It is the glue that sticks atoms to each other and gives matter its commonly observed properties, from a puffy marshmallow to a hard diamond. The electron has considerable mass compared to light, hundreds to thousand of times more than the red photon.

It also has a negative electric charge, whereas the photon is neutral. (We did say the photon's electric sensor oscillates to give the photon its frequency but this is a sensor that merely detects an electric charge. Humans can sense honey with their taste buds but this doesn't mean honey is produced by their tongues).

Why the electron has the specific properties it exhibits is unknown but, like the photon, it is a form used by the Creator to package energy. While the photon can have almost zero energy in the range of very long wavelength radio waves, the minimum energy for an electron is a half million eV, the energy equivalent of its mass, $E=mc^2$. When set in motion the electron can increase its total energy further, depending on the square of its velocity. Thus, if we venture to interpret the scheme of things, photons take care of bundles of energy from zero to about a million eV, above which electrons are a more efficient (or preferred) form of packaging.

To compound the mystery, nature has created an electron "sister" called the positron - the same mass as the electron only positively charged. But the positron is an ugly duckling - it is rarely seen. What it has to do with the grand scheme of things is not obvious, it's like a leper that nature acknowledges but prefers to hide.

To picture an electron think of a small ball of negative electric charge whose size is about one hundred millionth the wavelength of a red photon, which we've already said is one hundredth the diameter of a human hair. It is interesting that when the photon energy is increased to one million eV and it becomes eligible for conversion into an electron-positron pair, its wavelength has been reduced to about the size of an electron. We can venture a guess that this may be related to the reason for the changeover from photon to electron, but it's mere speculation.*

B - Properties

While the energy of an electron at rest is given by Einstein's equation, an interesting and unexpected thing happens when it starts moving - its mass increases. If you accurately measured the mass of an electron you'd know how fast it was moving, compliments of Albert Einstein. At one percent of the speed of light i.e. - 1860 miles per second, an electron (or a man) increases its weight by one hundreth of one percent, virtually immeasurable. (It's much simpler to measure its speed directly). At ten percent of the velocity of light the weight increases by 2% while at half the velocity we are 15% heavier. If you could manage to approach to within ninety percent of the velocity of light you'd more than double your weight. Only in special devices like synchrotrons and linear accelerators can scientists accelerate electrons this close to the speed of light. These high velocities are not achieved in vehicles that transport people, hence the mass increase is academic for we mortals.

So, no matter whether we determine the speed of an electron by clocking it between two points, or we measure its mass accurately, we are determining the same quantity - its energy. As far as we know the total negative charge on the electron is unaltered as we speed it up, so where does this extra energy go?

At about the same time that Compton demonstrated the momentum of the photon, a Frenchman, Louis de Broglie, postulated that an electron had a wavelength and this was later confirmed experimentally by scattering electrons from

---

* The observation that negative and positive charges in matter are attracted to each other while like charges repel is attributed to America's first physicist, Dr. Benjamin Franklin, who performed a series of experiments on static electricity, kite flying in stormy weather, and on lightening arrestors. He was awarded a doctorate from St. Andrews for this effort and was invited to become a member of the Royal Society of London in 1756.

crystals whose regular spacing between atoms created special conditions for the electrons to be diffracted and permitted one to ascertain the wavelength. The electron's wavelength might be associated with a pulsating electric charge, just as the photon's wavelength arises from the oscillating electric sensor. Thus, as the electron speeds up, its electric charge pulses more rapidly - a crude picture as to where the extra energy goes.

C - Particle or Wave?

The same dichotomy about wave or particle for the photon applies to the electron. Its mass suggests a particle while diffraction from a crystal indicates a wave character. The wavelength of an electron of energy equal to that of a red photon (1.6 eV) is about 1 nanometer or 700 times smaller than the photon's. As the energy increases the photon's wavelength decreases more rapidly than the electron. At 10,000 eV a photon's wavelength is about 0.12 nanometers and the electron's 0.012 nm, the two differing by a factor of 10. At one million eV the photon wavelength is 0.0012 nm and the electron 0.004 nm, the photon is now shorter.

The electron's wavelength must not be confused with the size of the electron itself, 0.00028 nm, representing the very heart of the mysterious forces that hold it together and produce the electric charge.

D - Spin

In the 1920's physics was undergoing a revolution with the introduction of quantum mechanics and, as part of its development, Goudsmit and Uhlenbeck, two Dutch physicists, postulated that the electron was spinning on an axis, a fact later substantiated experimentally. Just like a spinning top this produced an inherent angular momentum, and, since its electric charge rotated also, gave rise to a magnetic field with a north and south pole. This magnetic field accounts for the magnetism of the metals iron, cobalt, nickel and gadolinium. As far as we can tell the angular momentum and magnetic field decrease as the electron increases its linear speed, just as the wavelength changes.

By virtue of its mass the electron is subject to gravitational attraction, but this is considerably weaker than its electrical or magnetic forces. One of the hopes of theoretical physicists, and a problem that interested Einstein, was to provide a simple picture that would give rise to all three forces. At present the electric, magnetic, and gravitational forces are treated as separate entities.

It is interesting that in altering the energy package from photons to electrons the Creator made such a radical change in the properties. Furthermore, it is a puzzle (to me) why our universe consists primarily of electrons rather than positrons - the brother and sister pair being mirror images.

A crude picture of the electron is a small ball of pulsating negative electric charge spinning on its axis and producing a magnetic field. As it picks up linear speed, the electron's pulsations increase in frequency and this gives rise to its decreasing wavelength. The mass, hence gravitational attraction, of the electron is intimately tied up with its electric charge.

E - Emitting Photons

Electrons and photons are closely linked. For one thing they are modestly affected by each other's presence, unlike photons that ignore each other and electrons that strongly repel each other.

If a moving electron enters a magnetic field its electric charge causes it to swerve in its path and, in so doing, lose energy by shedding photons, a consequence of the conservation of momentum. Stated simply, if an electron swerves in one direction Newton's Third Law of Motion requires that an equal and opposite reaction must occur - hence, to avoid the wrath of Sir Isaac it emits photons.

If an electron and positron join up as a pair they soon annihilate each other (after they slow down) and create two photons of energy half a million eV (a process called positron annihilation), thus conserving the total energy. Positrons are doomed to extinction in our world for, straying too close to a sacrificial electron, the pair go out like a light. The reason is that the total positive and negative charges in the universe seem to just cancel - the Creator's insistence on good bookkeeping.

F - Beta Rays

We will have little to say about radioactivity but we will mention that certain nuclei reduce their energy by shedding an electron, historically known as a beta ray. The electron is not contained within the nucleus but is a convenient form for the nucleus to dispose of its energy. (Words come out of our mouths but that does not mean that words are stored there).

G - Other Particles

The Creator has produced a myriad of other particles, most of which are not relevant to our story, although two are germane to the conceptual pictures we are painting for the behavior of matter. The proton and the neutron are energy packages about 1850 times more massive than the electron, the former with a positive electric charge equal in magnitude to the positron and the latter without charge. Both are spinning on their axes producing a magnetic field significantly weaker than the electron's. The attraction between the proton's positive charge and the electron's negative charge is primarily responsible for the atomic structure of matter. The larger mass of the proton and neutron means that their minimum energy is almost one billion eV. Between the photon, the electron, the proton, and the neutron we can provide a good description of our natural world.

ELECTRON PROPERTIES

1. Minimum energy slightly over half a million eV when standing still. As it moves it increases its mass, hence its energy.
2. It has a negative electric charge that does not vary with energy.
3. Its electric charge oscillates with time and this gives rise to its wavelength, shortening as the velocity increases.
4. It is constantly spinning on its axis.
5. It is a small magnet.

6. It has dual wave and particle properties, just like a photon.
7. It has an "unwanted" sibling called a positron, oppositely charged but with the same mass, and spin.
8. It is attracted by gravitational forces.
9. Coupled with the more massive protons and neutrons, it holds atoms together in solids, liquids, and people.
10. Its size is small, 0.00028 nanometers.

NOMENCLATURE

1. Electron and positron have identical mass, and are oppositely charged (negative and positive).
2. Both have identical angular momenta and equal but opposite magnetic polarity.
3. Beta rays (either electrons or positrons) are shed by certain radioactive nuclei.
4. Positron annihilation - When a positron and electron come close to each other they soon annihilate creating two photons of half a million eV energy. The positive and negative charges disappear, just canceling each other.
5. Protons and neutrons provide nature with more massive energy packages than the electron.

CHAPTER VI

Early 20th Century
Part 2

Spectroscopes, better resolution gratings, and higher powered telescopes and microscopes occupied the attention of the world's small scientific community in the early twentieth century. The physical societies were flourishing although the man on the street scarcely know the meaning of the word physics, mostly confusing it as a cathartic. Yet each decade was presented with a multitude of discoveries that seemed comparable to centuries of development in bygone years.

Two revolutions burst on the scene. In art the New York Armory show of 1913 introduced the wild machinations of the Cubists, called crazy by several Presidents. It started in 1912 when two art experts visited a grand exhibition of modern works in Cologne, Germany. Overwhelmed by 125 van Goghs, 26 Cezannes, 25 Gaugins, and 16 by Picasso these two were spurred on to duplicate an exhibit of this magnitude in New York. Renting the 69th Street Armory, art studios, galleries, critics, and patrons were inspired to produce the greatest art show in America. It opened on February 17, 1913 to a patronage of 4000 to be followed in weeks to come by 70,000 visitors. Marcel Duchamps' NUDE DESCENDING A STAIRCASE, the most talked about work, was sold for the ridiculous price of $324, a Cezanne fetching $6700.

The publicity given this show had every New Yorker reacting either in shock or excitement, particularly about Duchamps' cubist painting which one critic described as an 'explosion in a shingle factory.' Teddy Roosevelt kept asking the question, "What is art?" The Armory show had represented a revolution in art as profound as quantum mechanics in physics.

In science the quantum theory elicited similar responses amongst physicists. It made no sense to men like Lord Rayleigh who argued that such explanations as quantization were as bad as no theory at all. Nonetheless the effort to explain scientific observations only made progress through such radical ideas. The man who introduced the quantum theory to the world in 1900, Max Planck, struggled hard to find the key.

Quantized energies merely mean that electrons on atoms are only comfortable in certain positions and with a certain range of momenta. They very rapidly accommodate themselves to their surroundings and it is only at certain energies that they can survive for long. When they jump from one energy to another photons of specific or quantized energies are emitted or absorbed. After a few years of living with the concept you begin to accept it. (In any event we believe the electrons are happy with it.) What about Max Planck who gave birth to the idea?

Creator of one of the fundamental breakthroughs in science and the father of modern physics, Max Planck was virtually self-educated in physics, although he received a good formal education. It required the same resolve that produced Einstein's theory of relativity and Newton's *Principia* to have been able to discard classical concepts in physics and introduce something 'crazy' like quantized energy levels.

Born in Kiel, Germany, in 1858, Planck's formative years were spent in a world of physics that appeared continuous and smoothly behaved, just as in our macro-scopic everyday world. The first experimental facts defying such a classical description were the wavelength distribution of radiation from a 'black body' and the low temperature specific heat of solids.

Planck's doctoral thesis on the second law of thermodynamics was produced with very little help from his thesis advisers. As Planck stated in his memoirs he decided that every step of his research had to be understandable to himself. He knew that substances heated as black bodies (i.e. they radiated light from a heated cavity) emitted spectra independent of the material out of which the black body was made. A black body is a hollow sphere with a small hole that permits photons to escape from the interior.

Physicists were able to produce experimental results that fitted both the low energy and high energy end of the spectrum, but did not know how to join them theoretically. To do so, Planck found that he had to resort to a quantized description of the energy with constant of proportionality h. In this way, Planck produced his Nobel Prize-winning energy distribution that fitted the entire spectrum. He reasoned that the black body radiation and oscillators inside the cavity were in equilibrium, but that the oscillator energies were quantized in integral multiples of h times the frequency. Walking in the Grunewald in Berlin when the idea came to him he turned to his son and said he had made a discovery as important as Newton's. Planck recounts his dilemma before the discovery:

I can recall the whole action as a process of despair. Actually, my nature is peace-loving and I am disinclined towards serious adventure. But for six years I had been doing battle with the problem of the equilibrium between matter and radiation without success. I knew the problem to be of fundamental importance to physics and I was familiar with the formula that reproduced the energy distribution in the normal spectrum. I was convinced that a theoretical interpretation would have to be found at any cost, no matter how high. Classical physics was not adequate, that was clear to me.

It was on 14 December 1900 that he appeared before the German Physical Society and presented his quantum of action, as he called it. It was an auspicious start to the century. Once quantization was accepted, theoretical physics was never the same.

Years later in reminiscing over those days he said:

This experience gave me an opportunity to learn a fact - a remarkable one, in my opinion. A new scientific truth does not triumph by convincing its opponents and making them see the light, but rather because its opponents eventually die, and a new generation grows up that is familiar with it.

Planck's personal life suffered from endless tragedy. His wife died in 1909 and he lost a son in the Great War. His two daughters died in childbirth at the time of that war, and a second son was implicated in the famous 20 July 1944 assassination attempt to kill Hitler and was tortured to death. Planck's home was bombed and his vast library and personal papers were destroyed. During one air raid he was buried alive for several hours.

Planck grew up with a strong sense of duty to the Fatherland, having employed this patriotic argument to persuade Einstein not to leave Berlin after the Great War. Yet when anti-Semitism reared its head again in 1931, Einstein drafted, but never mailed, the following letter to Planck, a man he revered:

> You will surely recall that after the war I declared my willingness to accept German citizenship, in addition to my Swiss citizenship. The events of recent days suggest that it is not advisable to maintain this situation. Therefore I should be grateful if you saw to it that my German citizenship were revoked, and to advise me whether such a change will permit (which I sincerely hope).

How it must have pained Planck to have to write a few years later, after Hitler came to power, that Einstein's resignation and emigration was an honorable decision and would spare Einstein's friends immeasurable grief. Caught up with the Nazi frenzy in 1933, Planck did not realize the tragedy that he and Germany would face in the next dozen years. Of the Old Guard that remained in Germany only von Laue was outspoken in his refusal to acquiesce to the regime's racial philosophy.

Planck believed that Nazi oppression would cease in due course. He remained constant to a Germany that failed its scientific tradition and destroyed the halcyon days of Göttingen, Heidelberg, Munich and Berlin. In his later years much of his life was spent in philosophical writing and scientific administration and he died a broken man in 1947.

The idea of quantized energies for the oscillators of each frequency in a black body cavity is totally alien to our everyday experience. The justification for this scheme arose mostly because it worked, deeper understanding was expected to come later. This, indeed, is typically the modus operandi of the physicist. One stays awake at night pondering a problem and relies on a vivid imagination if all rational ideas fail. To argue that the only oscillators that exist in a black body were those whose energies were $h\nu$, $2h\nu$, $3h\nu$, $4h\nu$, etc. is hard to swallow but a certain amount of post mortem examination of the classical theory of optics might help one accept this approach.

Five years after Planck, Einstein introduced his revolutionary view that light itself behaves like quanta (particles, now called photons) under certain circumstances.

Using this concept, the argument then proceeds as follows: If one places these quantized photons in a box, their wave character permits them to avoid interference with each other, hv has nodes at the opposite walls of the box 2hv has nodes at the walls and one in the middle, 3hv has notes at the walls and two within the box, etc. The positive and negative parts of the waves of differing photons cancel each other so that the photons do not disturb each other. This is a wishy-washy physical picture that may satisfy you enough to go on to the next paragraph. If not, trade this book in on a new bicycle and commune with nature or spend four years in graduate school.

By utilizing this quantum hypothesis, Planck produced a theory that behaved as the Creator ordained.

Once quantization took root Einstein in Switzerland, Bohr in Denmark and Born in Germany, plucked it from the garden and planted its seeds amongst the younger generation of physicists, still an exclusive society of clever students some of whom lived at poverty scale but all of whom were driven with a desire as great as an artist living in the proverbial garret. This group of theoretical physicists may have been considered just as crazy as modern painters although they did not share the publicity of Picasso, Braque, and other modernists. At least the physicist was intent on explaining nature's behavior - who knew to what higher authority the modern artist was beckoned? Besides, when Picasso paintings in 1990 consistently sold for 10 to 50 million dollars the world had long ceased asking the question.

The revolution in art occurred at the same time as that in physics but the two are scarcely related. Picasso probably cared little for the work of scientists. There have been several biographies about Picasso whose prodigious output of at least 10^5 paintings, drawings, doodlings on tablecloths, and ceramics makes cataloguing an impossible task - he is a popular target for the forger. Even a signed painting comes into serious question without reliable provenance. The verbiage expended by art writers, historians, and critics in attempting to unlock the secrets of Picasso's genius is all too benumbing. If one wishes to understand his art you must look at the thousands of his paintings, digest the overall impression and then look at them again. Whether this process is worth your while or whether it will rival the success of quantum theory by an insightful interpretation of man's psychological response to shape and color is anyone's guess.*

Unlike quantum theory which attempts to guide us in the footsteps of the Creator when submicroscopic entities are examined, artists like Picasso are creating their own interpretation of the man-sized world, with rules that appear beyond comprehension by the written word. We are continually stimulated in our world by images with an infinity of shapes and color arrangements; light drives us toward the pleasures of aesthetics; and we realize that we have evolved with an innate desire to explore its many ramifications. Even the Impressionists dabbled in

---

* Picasso quote: When I was a child my mother told me, 'If you become a soldier, you'll be a general; if you become a monk, you'll end up as the Pope.' Instead, I became a painter and ended up as Picasso.

a kind of quantized scheme for color application but they scarcely engaged in a dialogue with the physics community. The physicist's task in deciphering the world of the submicroscopic was simple because all atoms followed certain rules. The artist lives in a world virtually devoid of rules - the possibilities are surely infinite.

CHAPTER VII

Tutorial 2
Photons and Electrons

If it were not for electrons, photons would live in a dreary world of nothingness. Moving about aimlessly, it would be like a day without sunshine. The photon and electron were meant for each other and their love affair enables us to employ the two in the enrichment of art, and technology. Let's examine how these two creatures live together.

A - Their Interaction

Much of our picture of photons and electrons emerges from observations of how the two behave toward each other, the primary interaction coming from the photon's electric sensor feeling the charge on the electron. If we shoot high-energy gamma rays at a thin foil of matter we find that some fraction of the gamma rays pass through unaltered in energy or direction, since much of matter is empty space, while the remainder either bounce off in various directions (called scattering) or are gobbled up and disappear (called absorption). A considerable history of research surrounds this simple experiment but from its results we can determine some fundamental properties of matter such as the size of the electron, how the electrons (hence the atoms) are distributed in the foil, and even perform a chemical analysis of the foil. By replacing the foil with a person we can perform X-ray diagnostic measurements such as looking for broken bones or cavities in teeth, or, alternatively, perform therapeutic work such as destroying tumors. By substituting the human with a turbine blade, we can look for cracks and strains in the metal. Shooting high energy photons at all manner of things has developed into a large industry, the research journals publishing thousands of papers a year.

For the most part physicists today are satisfied that they understand the manner in which photons and electrons respond to each other. But when Arthur Compton did his experiment in the early 1920's he came up against a puzzle, a frequent occurrence that precedes discoveries in physics. He started with a beam of X-rays of a single energy of about 17,000 eV and directed them at a piece of graphite, analyzing the X rays that bounced back. To his surprise they had lost several percent of their energy. Since the mass of the electron had already been measured years before, Compton was led to conclude that the photon possessed a momentum given by its energy divided by its velocity (the speed of light) and the photon and electron had collided like billiard balls of differing mass. Unfortunately the electron did not pick up enough energy to escape from the graphite so Compton had to infer the sequence of events solely from the energy and direction of the scattered photon. But he was able to show that as the scattered angle was decreased the energy loss of the photon was reduced in a way expected for billiard balls. (If you measure a cue ball's direction and speed both before and after a billiard shot, you can calculate the direction and speed of the ball that it has struck providing you know the later's mass).

Six years after Compton's measurements J.W. DuMond in California improved the experimental accuracy of the work and showed that the X-rays scattered in any particular direction had a range of energy losses, evidence that the electrons must have been moving before the collision. Between the time of Compton's discovery in 1923 and DuMond's results in 1929 quantum mechanics had burst on the scene and it provided scientists with an understanding of the electron speeds expected in graphite. Since that time, the Compton experiment has been performed on a large variety of substances enabling us to expand our understanding of the electron behavior of matter and from this, the nature of the chemical bond that holds atoms together.

The Compton experiment ranks as the most fundamental in our understanding of the properties of the four most common energy packages, photons, electrons, neutrons, and protons since the experiment involves a single photon scattered by a single electron. There have been other scattering experiments like electron on electron, electron on proton, proton on proton, neutron on electron and neutron on proton but none have the elegance and simplicity of interpretation of the Compton results, although the electron-electron scattering experiments have been very instructive and are the basis for the electron microscope.

B- Relativity

Einstein's special theory of relativity has been expounded and interpreted in a plethora of publications so we only need underscore its most obvious result, i.e. the masses of electrons, protons, neutrons, and space ships increase significantly as they approach the velocity of light. Such high velocities are regularly achieved for electrons and protons in man-made accelerators like synchrotrons, and these high speed particles have then been employed to investigate the inner structure of the nuclei of atoms. The synchrotron has also been adapted to provide copious sources of photons by sending the high-velocity electrons on curved paths through magnetic fields that prod them to emit photons polarized in a plane tangent to their line of flight. The energy of these photons is in the range from 100 eV to thousands of eV, i.e., in the relatively weak X-ray region, but they are so intense in number that they make feasible studies of matter that can not be accomplished with ordinary X-ray machines.

While relativity increases the mass of the electron as it approaches the speed of light, it does not increase its size, rather its size and wavelength both decrease. The principal reason for introducing the subject of relativity is that it is one of the triumphs of man's exploration into physics, although it is not crucial to the understanding of the common properties of matter. The history of the events leading to Einstein's theory are still unclear since even Einstein appears uncertain as to his train of thoughts around 1900 when he was thinking about the question.

In looking for the aether Michelson wrote:

'To suppose that one body may act on another at a distance through a vacuum without the mediation of anything else, is to me so great an absurdity that I believe no man, who has in physics a complete faculty for thinking, can ever do.' Thus he joined Maxwell, Newton, Franklin, and others in espousing the aether.

When a disappointed Michelson failed to detect an aether, Einstein took the bold step of proposing that the aether did not exist and that the velocity of light was an upper limit to the speed of matter. In one fell swoop the photon became a particle totally independent of an aether. Einstein developed his ideas by asking what might occur if there were two moving bodies (like spaceships) capable of measuring each other's speed and dimensions with photons, just as the police employ radar to nab speeders. By stipulating that energy was conserved and that the speed of light was sacrosanct, Einstein developed his famous equations such as $E=mc^2$ and the equations describing the increase in mass and decrease in dimensions of the spaceships as the relative speed between the two approach the velocity of light.

In our everyday world we rarely exceed speeds of one mile a second so we are not familiar with things that change their weight or their dimensions while moving. But enter the Alice in Wonderland treadmill of high velocities and we have difficulty adjusting our imagery.

Can we carry a simple picture in our mind to allow for changes in the Creator's universe as we approach the speed of light? If we remember that the mass, hence the energy, of the electron increases; that its wavelength decreases because its oscillating electric field is increasing (the one that defines its wavelength); that the shortened wavelength will bring atoms closer together and shorten the dimensions of matter; and that the increased mass makes electrons more sluggish and slows down their movement - hence their time scale - we have a crude picture of events on the treadmill of Wonderland. Finally, if we postulate that energy is unchanged from the treadmill world to the stationary world, i.e. a ball tossed by Alice from the treadmill and caught by the Mad Hatter before it breaks any teacups has the same energy, then Einstein's equation $E=mc^2$ emerges.

C - Atoms and Energy Levels

We shall utilize the properties of the four energy packages - photons, electrons, protons, and neutrons - to build up the world around us, the everyday world that is not subject to those unexpected events of high energy or high speed. The hydrogen atom is the simplest package that we can deal with, and we can trace the same path of historical curiosity that led physicists to the discovery of quantum mechanics in 1926. For this story we need the electron, the proton, and the photon.

We begin with a negatively charged electron and a positively charged proton, the latter 1847 times heavier but with the same magnitude of charge. Place this combination someplace they will not be disturbed, like in the vacuum of outer space, and you have a stable hydrogen atom, less than a nanometer in size (divide a meter stick into 10^9 equal parts). The electron is engaged in maneuvers around the proton, sometimes going one way then another. It is in the nature of the Creator's world that He keeps the details of this movement from us - neither quantum mechanics nor any experiment we can perform provides us a detailed answer. We can only determine the probability of finding the electron at various distances from the proton and the probability of it having certain values of velocity. We are led to assume that the electron must move over some path to achieve these positions and velocities. The range of distances covers about a nanometer, ten thousand times larger than the size of the electron - something akin to the size of the earth compared to its distance from the sun - i.e. a hydrogen atom is primarily empty space. The electron speeds cover a range from nearly zero to about 10^5 cm/sec - i.e. somewhat faster than a high-speed bullet.

But unlike the earth orbiting around the sun at almost a fixed distance and a fixed speed, the electron sometimes approaches very close to the proton and at other times is so far away as to appear to escape from the proton. In its gyrations around the proton it changes its speed, very high when close to the proton and small when far away. The attractive electric force between the electron and proton varies as the inverse square of the distance and the electron compensates for this variation by continually altering its velocity, so that the sum of its positive kinetic energy and negative electrical energy of attraction remains constant.

In such a benign state of affairs physicists learn little about the properties of the hydrogen atom. One must disturb it to glean information. The observations that first interested and perplexed physicists for over a hundred years was the behavior of the atom when excited in a gas discharge, effectively stripping the proton of its electron. When the same or another electron returned to the proton some time later, it would select one of a number of quantized or discrete energies, emitting a photon of energy equal to the difference of the energy of the benign (ground) state and the excited state. It was the discrete energies of these photons that baffled physicists. The electron only had a choice of specific orbits around the proton corresponding to very definite energy levels. How come?

In developing quantum mechanics through the so-called Schrödinger equation, physicists were able to calculate these energy levels for the hydrogen atom provided they adopted certain reasonable rules for the mathematical solutions to the equation. By so doing, both the photons emitted as the atom passed from one energy state to another and the photons that could be absorbed between energy states were clearly identified by the theoretical calculations and confirmed by the experimental observations.

Why only specific quantized energies? The electron had to restrict its motion to paths that kept the energy constant for each state; the different states could not mathematically interfere with each other lest the atom become unstable; and the electron could not wander too far away from the proton lest it be gone forever.

These conditions only permitted specific orbits but with these rules the physicist could produce exact agreement with the hydrogen atom - a triumph for quantum mechanics and a Nobel Prize for Schrödinger and Heisenberg (the latter, working with Max Born developed quantum mechanics in a different mathematical form). Jubilation reigned amongst the physicists and chemists!

Alas! The exultation was short-lived because the simple hydrogen atom was the only one the Schrödinger equation could solve. The next more difficult atom was helium with two electrons, and a nucleus containing two protons and two neutrons, the neutrons uncharged and only adding mass to the nucleus. The problem arose because the electrons were negatively charged and repelled each other. This repulsion between electrons has since been the bane of physicists engaged in calculating the electronic energies and electron orbits of atoms using the Schrödinger equation.

Theoretical physicists have settled for a compromise: rather than demand an exact mathematical answer they have found ways to manipulate the Schrödinger equation to give approximate, but reasonably accurate, answers. When experimentalists have employed X-rays to measure the energies, positions and velocities of electrons on atoms from helium to uranium they could do so with about the same accuracy as the theoretician did his calculations (0.1% error for the energies and 1% for the positions and velocities). It provided the theorists and experimentalists with a competitive incentive - if they didn't agree they could blame the error on each other.

D - Molecules and Solids

While the electron structure of atoms is fairly well understood, it is the buildup of atoms to form molecules, solids, and people that has created a major challenge in science and presents us with significant problems in measurement and understanding. The simplest molecule is hydrogen. Bring two atoms of hydrogen together and they will form a stable molecule. One may ask why this happens since the electric charges of the electron and the proton just cancel and the two atoms should be unawares of each others' presence.

But if we examine the configuration of two electrons and two protons and add up the repulsive and attractive energies (the forces vary inversely as the square of the distances and the potential energies inversely as the distance), and take an average for all values the attractive (negative) energies come out ahead. In the hydrogen molecule (as in all matter) the electrons take up a whole range of favorable positions as they interweave their orbits around the sluggish heavier protons that remain relatively fixed in position, content to vibrate short distances along a line connecting them. This is the nature of the chemical bond holding matter together.

When the photons are brought into the picture of the hydrogen molecule, we find that the quantized energy levels of the hydrogen atom are totally altered in character. For one thing we encounter a set of closely spaced quantized levels that are associated with the two vibrating protons and an even more closely spaced set of energy levels associated with the protons rotating around an axis perpendicular to the line joining the protons, like a dumbbell. These are called vibrational and rotational levels respectively and they will absorb or emit a wide spectrum of photons as they jump from one quantized level to another.

If the molecules of hydrogen are brought together, as in a tank of hydrogen gas, they engage in a number of machinations. In addition to their vibrations and rotations they move bodily through the gas, colliding with other molecules and often bouncing off the walls of the tank. When they collide with each other they can swap energy by changing their rotational levels and speeds.

Gases are one common form of matter although we are equally familiar with liquids and solids. But liquid and solid forms of hydrogen only occur at very low temperatures since the hydrogen molecules are weakly bound.

There are many energetically favorable arrangements for atoms and molecules that produce such varied forms of matter as metals, plastics, organic materials, hamburgers, ceramics and puppy dogs. Unfortunately there are no rules to tell us what will emerge when various elements are combined and almost all of our technological achievements with materials have resulted from trial and error, the Schrödinger equation being woefully inadequate in predicting the properties of matter.

Although studies of the electron arrangements that bind atoms together have been pursued by physicists and chemists, the vast bulk of the scientific community overlooks the precise role played by the electrons and concentrates on the arrangement of the atoms themselves. A considerable fraction of solid matter is crystalline, the atoms lining up in a regular three-dimensional array. Such substances are studied by crystallographers employing X-ray diffraction, a technique utilizing photons whose wavelength is comparable to the spacing between atoms. The pattern of scattered X-rays is unique for each crystalline arrangement so that a diffraction fingerprint is formed. However, there are just as many solid substances that are not crystalline (or only partially crystalline) and we are less certain about their atomic arrangements. A short list of these non-crystalline substances includes wood, skin, plastics, glass, rubber, corn flakes, rugs, people, and escargots. In short, much of our organic world is non-crystalline and biochemists often have to use other less precise measurements to provide clues to the structure of such solids. (When the complex structure of DNA was solved a crystalline sample had to be produced by Rosalind Franklin in order to provide the X-ray fingerprints).

E - Photons and Solids

One of the technological achievements of this decade is the optical fiber , a high purity non-crystalline form of glass (SiO_2) drawn into hairlike fibers miles long and capable of transmitting visible light over these distances - a kind of optical "superconductor". This is accomplished by ensuring that the glass is free from those impurities that can either absorb or scatter the photons and by coating the glass with a cladding that acts as a barrier to prevent photons from escaping as the glass is subjected to slight bends along its length. The cladding is a specially treated glass with elements added to reduce its index of refraction by about 1% from the core. Light that strikes the core-cladding boundary at low angles bounces back into the core just as light from air is reflected back from a water surface when the angle is less than about 55 degrees (Brewster's angle).

We are familiar with a wide variety of solids that respond differently to light because the electrons that give the solid its particular properties - metals, ceramics, dyes, glasses, superconductors, semiconductors - have different quantized energy levels. As we've already hinted, these energy levels are beyond the ken of the theoretician and quantum mechanical calculations. There are too many electrons, too many energy levels, and too many electron-photon possibilities to be predicted by the theoretician. Fortunately we can generally measure these levels and save the theoretician any embarrassment. In contrast the theoretician can produce models of these complex structures that yield qualitative answers to the quantized levels.

Metals both absorb and reflect visible photons, although the more energetic photons like X-rays penetrate most materials; ceramics such as sapphire, diamond, rutile, silica, boron oxide, etc. are transparent since visible photons do not have enough energy to excite electrons to higher energy quantized states; dyes contain elements or compounds with energy levels in the visible range of light energies, around one to two eV, hence they select specific photon energies for absorption and scattering; semiconductors contain electrons that absorb all photons above a certain energy range, called the band gap, after which these electrons behave as in metals; below their critical temperatures superconductors absorb photons like a metal. The subject of photon interactions with solids is very complex and our libraries are bulging with research publications on the subject. It is sufficient for our needs to merely point this out - the possibilities are limitless and new materials are reported annually. Hence research is fun because you might have a new material named after you.

F - The Role of Mathematics in Science

The theoretical physicist employs mathematical manipulation to obtain answers to whatever question he chooses to pose, like a new approach to the subject, a calculation of the properties of a substance, or an empirical representation of the behavior of matter. Whatever problem he selects, he works within the limitations of his calculator and his ability, but his calculations cannot be perfect. A few examples will help to illustrate this.

Twenty years after the discovery of the Nobel prizewinning Schrödinger equation that provided "exact" answers to the energies of the simple hydrogen atom (one electron, one proton), Professor Willis Lamb, Jr. working at Columbia University determined with high accuracy a discrepancy in a few of the energy levels, already hinted at by some previous work by Williams at King's College London. This fundamental and unexpected discovery earned Lamb the Nobel Prize in 1955 and forced the theoretical physicists to reassess the accuracy of the Schrödinger equation. They eventually attributed the "Lamb shift" to a small modification in the magnetic structure of the electron.

When Bardeen, Cooper, and Schrieffer were awarded the Nobel Prize for their theory of superconductivity, no on questioned their achievement. It relied on a model for a superconductor that had a gap in energy that separated the superconducting state from the normal state. But when the totally unexpected high-temperature superconductors were discovered 15 years later Bardeen admitted that his prizewinning theory could not explain the effect. At the present time the theoreticians are still working on this one, perplexed as to why superconductors may require two different theories. (Are the electrons confused or only the scientists?)

When Michelson failed to discover an aether in 1887, this perhaps contributed to Einstein's thinking about the subject and found its explanation in an important theoretical development in physics - relativity. Michelson received the Nobel Prize in 1907 for measuring the velocity of light and Einstein in 1921 for the photoelectric effect. Neither were awarded the prize for the work that produced relativity! Ironically, the very year Einstein published his famous paper on relativity (1905) the future ardent Nazi Lenard received the Nobel Prize for his work on electron beams. During the Third Reich and until his death in 1947 Lenard was an outspoken Aryan and critic of relativity.

A dozen years after his work on relativity Einstein proposed the phenomenon of stimulated emission, a theory that lay dormant until 1960 when it became the fundamental idea behind the most important technical discovery since relativity, the laser by T. Maiman. Since Einstein had died a few years prior, he was not eligible for a second prize but neither has Maiman been honored. Hence the laser and relativity have not been specifically recognized by the Nobel Committee although some work at much lower energies by Townes, Basov, and Prochorov did get the prize in 1964 for demonstrating stimulated emission.

When a Harvard theoretician created a new mathematics to deal with a specific problem in physics, he discovered that it was not applicable to the problem for which it had been developed. He searched about and fortunately found another problem in physics to which it could be applied. (Lucky soul - this does not happen often).

It is well to remember that theoreticians develop simple models to replace the real ones encountered in nature. Only in this way can he work with manageable equations. But suppose his model is a poor one? He may obtain answers close to experimental ones, but was that luck, intuition, or, did he stop his calculations when he got the known experimental answer? One must not denigrate the theoretician - but like the piano player in the wild west saloon he's doing the best he can, even though he's not certain what's going on upstairs.

Perhaps a simple example from our lives might illustrate the point. Suppose you lost your job as a mathematician and were hired to run a limousine company between two towns. In the interests of efficiency you wanted to calculate how long the journey would take. If you knew the approximate distance and the speed limit you could make a quick estimate but, if you wished to improve on that, you might add the fact that there were 22 or 23 stop lights along the way, although you didn't know how long the red and green alternated nor did you know whether any group of lights were staggered. You could start with a simple model in which all stop lights were 45 seconds long and worked independently, and calculate a time, but as the dispatcher began to record the experimental times for the journey, you could try to modify your model to include staggering of the lights, the actual on-off times, traffic conditions etc., ultimately hoping to achieve a better answer.

The theoretician employing mathematics as his tool faces these sorts of problems but, in addition, further considerations are necessary when new experimental results and new discoveries emerge, sometimes so rapidly that he has little time to keep up with developments. In the early part of this century a philosophy pervaded the scientific community that theoretical physics would provide the clue to man's understanding of the hand of the Creator. We have since learned that the Creator is always way ahead, and that every time the theoretician thinks he has solved a problem, ten new ones rear their heads. Nonetheless, one theoretician per decade may produce an insightful discovery, like the Schrödinger equation, relativity, or stimulated emission, which makes up for all the theoretical work that has taken the wrong path and ended in a cul-de-sac.

CHAPTER VIII

Early 20th Century
Part 3

A famous pioneer in physics during the early days of quantum mechanics was Max Born who, like Bohr, contributed equally as teacher and researcher. His story has its dramatic moments:

Shortly after Hitler's rise to power in 1933 Professor Max Born was dismissed from his post in Göttingen, based on a Nazi edict that had recently suspended 39 other distinguished Jews. Not long afterwards he received a letter from Professor Werner Heisenberg, recent Nobel Laureate for his discovery of quantum mechanics. To avoid possible interception of this letter sent to Born who had fled to Cambridge, England, Heisenberg posted it in neutral Switzerland:

Dear Herr Born,   Zurich, 25 Nov. 1933

If I have not written to you for such a long time, and have not thanked you for your congratulations, it was partly because of my rather bad conscience with respect to you. The fact that I am to receive the Nobel Prize alone, for work done in Göttingen in collaboration - you, Jordan, and I - this fact depresses me and I hardly know what to write to you. I am, of course, glad that our common efforts are now appreciated, and I enjoy the recollection of the beautiful time of our collaboration. I also believe that all good physicists know how great was your and Jordan's contribution to the structure of quantum mechanics - and this remains unchanged by a wrong decision from outside. Yet I myself can do nothing but thank you again for all the fine collaboration, and I feel a little ashamed.

With kind regards,
yours,
W. Heisenberg

Such a letter was not easy to write but it did reflect that Heisenberg's mentor at Göttingen had made a significant contribution to the work that gained the young Heisenberg the greatest recognition in physics. The embarrassment reflects the awkward position for Heisenberg who did not share the Nazi obsession with anti-Semitism but decided not to abandon his country. He hoped that the Nazi regime would someday be toppled and he could help rebuild German science. His decision to stay in Germany contrasts another Nobel Prize winner for quantum mechanics, Erwin Schröedinger, an Aryan and Austrian who refused to have anything to do with the Nazis and left Germany already in 1933.

Max Born's contributions to optics and optical levels is well known. He wrote one of the standard texts in the field and proposed one of the fundamental approaches to the scattering of light and X-rays, the 'Born Approximation'. At 72, and in recognition of this work (and the oversight in 1933), Max Born was awarded the Nobel Prize in 1954. Born turned out one of the most prestigious arrays of students that physics has known.

This was not the only time Born's contribution was overlooked and had to be acknowledged years later. Mott and Massey were responsible for such an oversight in their famous book on *Atomic Collisions*, and Mott apologized in the introduction he wrote to Born's biography 39 years later.

Max Born was born in Breslau, Germany, in 1882 and studied physics during those days before quantum mechanics when their ranks were small and support came from private individuals who were in turn rewarded with honorary degrees. He studied in Breslau and Göttingen and after two stints in the military became a lecturer at Göttingen. Following a visit of the famous Michelson, Born accepted a return invitation and visited Chicago. He continued on to California where he became violently ill on some local grapes but Gilbert Lewis of Berkeley took him to a restaurant in San Francisco where he ordered some 'tea' that cured him. It was pure Scotch, in violation of Prohibition.

On return to Göttingen, Born shared a flat with the aerodynamics expert von Karman. The landlady had advertised for tenants suffering from mental disorders, but von Karman had convinced her that theoretical physicists needed just as much care. Using a combination of their names, they called the establishment Bokarebo. In 1913 Born married a clergyman's daughter and took Baptism. He never practiced formal Judaism.

In 1914 Born and the younger Ewald did some of the earliest work on infra-red properties of crystals and crystalline vibrations. On the invitation of Max Planck, Born left for Berlin in 1915 where he befriended Einstein and enjoyed piano and violin duets with him. As the war began to strain the economy, Born joined the Army group concerned with developing radio and sonic direction finders for locating enemy guns and also suggested the use of sonar for locating submarines. He found himself with spare time in the Army and did some theoretical work on optical polarization in non-centrosymmetric crystals like Iceland spar (non-centrosymmetric means that the mirror image of the atomic structure differs from the original structure). While still a sergeant he joined a committee to advise the Kaiser on unrestricted submarine warfare.

Germany was decimated at the end of the Great War and Born returned to a country of rioting, inflation and political uncertainty. He formed a research group in Frankfurt and, to counter the rampaging inflation, he invited his now famous friend Einstein to lecture and charged an admission fee. He also befriended the

American banker H. Goldman who provided him financial support.

In 1920 the first serious signs of anti-Semitism in the sciences emerged at a scientific meeting in Bad Mauheim, when Einstein's theory of relativity came under fire from two Nobel Laureates, Lenard and Stark. The meeting was attended by Einstein, Planck, Born and most of the outstanding German scientists of the day. Planck used his authority to prevent an open confrontation and Einstein listened with some amusement and did not bother to rebut the criticism.

In 1921 Born returned to Göttingen and over the next dozen years the elite of physics came to study there, including Pauli, Heisenberg, Jordan, Hund, Heitler, Condon, Oppenheimer, Fermi, Dirac, Hylleras, von Neumann, Wigner, Teller and Fock. The revolution that led to the development of quantum mechanics partly resulted from the thesis work of de Broglie in France who suggested that electrons might possess a wave character. The Born, Jordan, Heisenberg group developed the matrix formulation of quantum mechanics and this was successful in predicting the optical levels of the hydrogen atom, a problem earlier attacked by Bohr in 1913.

Born and Oppenheimer developed an important concept that is extensively used in quantum mechanics to separately treat the electrons and the nucleus of an atom. Born also developed an important idea for the scattering of photons and electrons from matter. Called the 'Born Approximation', it is a simplification that assumes the scattered wave escapes before it is scattered again. When applicable it simplifies calculations.

Oppenheimer, the Wunderkind from America, who came from well-to-do parents was considered obnoxious for his lack of continental manners. During some of Born's lectures, Oppenheimer would interrupt, take the chalk out of Born's hand, and proceed to suggest a better way to attack the problem. Born's students objected and threatened to quit the class unless something was done. By subterfuge, Born arranged for Oppenheimer to discover the student letter of protest and the problem disappeared.

With the suspension of the 40 professors in early 1933 and the excesses of the Nazis, Born burned some of his controversial personal library including *Das Kapital*. A chance meeting with Albert Schweitzer playing an organ in a German church, prompted Born to later write to him in Africa asking that he return to become an anti-Nazi spokesman. Schweitzer rejected the invitation saying Germany was a lost cause. Planck stuck his neck out at this time and arranged an audience with Hitler to plead the cause of the dismissed Jewish scientists. The famous physicist had to sit through a long tirade against the Jews and left totally dejected.

Born temporarily moved to Cambridge, England, where he was visited by Heisenberg with an offer to return to Germany and direct research at Göttingen provided he did not teach. That this undertaking did not include his family was the crudest of insults and Born asked Heisenberg how he could allow himself to bring such an unacceptable offer. Heisenberg did not approve of the Nazis but, like others, permitted himself to be swayed by excessive nationalism. It was difficult to remain in Germany without adopting the party line, although Nobel Laureate von Laue did just that. After a brief period in Cambridge, Born took a Chair in Edinburgh where he remained for 17 years. His most famous student in Scotland was Klaus Fuchs, before Fuchs turned to spying!

Born was a pacifist at heart and was spared any involvement in the Manhattan project. He never received an invitation to join in the work since it was probably realized that his political leanings would make him less than enthusiastic. His student Klaus Fuchs did go on to make some unwanted contributions.

The first two decades of the twentieth century found Western society engaged in several wars, the Spanish-American, the Boer, and the Great War. But scientists were able to communicate with each other during these conflicts, even though on opposite sides. Pure science provided a bond between them that transcended national boundaries.

During the first decade of the new century several giant leaps in optics came from an obscure Swiss civil servant, Albert Einstein.

When Albert and Elsa Einstein visited the new observatory on Mt. Wilson in California, Albert was invited to peer through the huge telescope. His wife Elsa asked one of the scientists what they were looking at. 'They are studying the structure of the universe.' Elsa retorted, 'Funny, my husband does that on the back of an envelope.'

It is difficult to separate legend from reality for this ill-kempt genius whose absent-mindedness and disregard for the unessentials of life contrasted sharply with the sophistication and neatness of his scientific ideas. His personal needs consisted of a pipe, a violin, a sailboat, women and his thoughts; the rest were superfluous. Yet his description of the universe brought him unwanted accolades and he became something of a zoo-like spectacle, to be seen and admired but at a safe distance.

The world about him and the foibles of politics and social behavior mattered little. When his first wife returned from Serbia after a holiday with his boys and announced that they had converted to Catholicism, he was totally undismayed. He was happy in his sailboat, even during a lull in the wind when he'd bring out his notebook and go to work. Sailing also provided the privacy that fame denied him. Einstein, like Newton and Maxwell, best communicated with himself.

His youth was uneventful and gave no indication of his future accomplishments. Born in Ulm, Germany, on 14 March 1879, he moved to Munich, to Italy and then to Zurich where he was educated at the Polytechnic. Not until he married, had children, and took employment at the Swiss Patent Office in Bern did he have the opportunity to prove his mettle. While his job required him to rewrite patent applications and make them intelligible, it was the spare time afforded him that he cherished. From 1902 until his famous publications in 1905, the world of physics fascinated him. He had virtually no interactions with real physicists except through the learned journals yet he solved some of the outstanding problems of the time, not to say opening new vistas. It is conceivable that his genius was stimulated by his daily attempt to reduce each patent idea to its fundamentals.

In a single issue of the *Annalen der Physik* of 1905, now a collector's item, Einstein created theories from fundamental principles that brought an understanding to Brownian motion, the photoelectric effect, and the lack of an aether drift in the Michelson-Morley experiment. The statistical approach to Brownian movement provided a value to molecular dimensions, while his revolutionary light - quantum hypothesis, derived from fundamental statistical arguments, provided an explanation of the photoelectric effect and other phenomena that indicated that light could behave like particles and transfer all of its energy to an electron. This strained the concept of light waves and diffraction, and for years the wave and particle natures of light was taught as a dichotomy. But it was the special theory of relativity that led to the famous $E=mc^2$ and attracted the attention of the world of physics. The lack of an aether drift in the Michelson-Morley experiment had been rendered superfluous: light was self-contained and did not need a transmitting medium.

Only after Planck in Berlin took note of the article on relativity, was the scientific community's attention drawn to this obscure civil servant. Who was he? Laue, later to become a Nobel Laureate for his discovery of X-ray diffraction, travelled to Bern to investigate. Arriving unannounced, he was hypnotized by the youth of the man who came to the waiting room to meet him and, certain there must have been a mistake, permitted Einstein to return to his desk without greeting him. Later, he waited for Einstein after he left work and the two went for a long walk. During the stroll Einstein offered Laue a cigar which he found so foul-tasting that he permitted it to fall into the river as they crossed a bridge. Favorably impressed, Laue reported this back to the center of physics in Berlin, but cautioned against accepting cigars from the man.

Einstein next turned his attention to the low temperature specific heat of solids. Employing Planck's quantized energy levels, he blazed a trail for future developments. The quantized energy levels for atomic vibrations in solids have come to be known as phonons, since their energy is h times the frequency, as for photons, and their momentum is the energy divided by their velocity (the speed of sound in the solid). This extended quantized energy levels to heat motion in solids. Einstein's fame grew rapidly and he received an offer of an honorary doctorate at the University of Geneva, but believing it to be junk mail discarded it without reading. Only a follow-up invitation attracted his attention. At an award banquet at which he shared the honors with Madame Curie, he remarked to her that Calvin would have considered the sumptuous repast sacrilegious.

That same year he accepted the offer of a professorship in Zurich. His superiors at the patent office, unaware of his talent, could not fathom his reason for leaving and offered him a raise, which he rejected. Neither job was lucrative and Einstein would joke that he planted clocks all over the universe to expound his theory of relativity, but he couldn't afford one for his own bedroom. Within a short period of time his fetish for clarifying concepts and his sense of humor, made him a popular lecturer at the university. His lucid explanations about relativity evoked his comment that the mathematicians at Göttingen ravelled his theory into complicated expressions merely to show how clever they were.

Einstein was courted by scientific establishments for the remainder of his life.

Within a few years he moved to a better position in Prague, then part of the Austro-Hungarian Empire. As a non-citizen he was granted a special dispensation by Emperor Franz Josef, a man friendly to the Jews. While in Prague, Einstein produced a paper that pointed out that light from a distant star would be bent approximately one second of arc as it passed the sun's gravitational field. Only during an eclipse could the starlight be seen against the sun's brilliance. This prediction would later bring him world fame.

The first Solvay Congress in 1911 consisted of an invited group of distinguished physicists, housed in the exclusive Hotel Metropole in Brussels. Einstein delivered a paper on specific heats, quantum theory occupying the forum's attention. When the great Rutherford was told that no Anglo Saxon understood relativity, he replied they had too much sense. Michelson never accepted it, even though it explained his null aether drift experiment.

In 1912 Einstein was enticed back to the Polytechnic in Zurich, partly because the opportunity for sailing was better than in Prague. Why was he so restless? The answer is not clear but we cannot rule out pressure from his wife. The marriage was in trouble and his wife was not happy in Prague. It was simpler for Einstein to move than to endure her nagging. Only a year later he was lured by a personal invitation of the Kaiser to head the new Kaiser Wilhelm Institute in Berlin and he may have employed this ruse to accelerate the breakup of the marriage. Besides, there were good opportunities to sail in the Berlin area.

Kaiser Wilhelm II, cousin to King George V and grandson of Queen Victoria, was under compulsion to make Germany first in virtually everything. His build-up of the navy brought him into direct conflict with the English who considered the seas their domain of responsibility. Kaiser 'Bill' was a somewhat irascible figure who took a personal interest in technological matters. The Kaiser Wilhelm Institutes, one for physics and one for chemistry, were privately financed by industrialists who were given special titles, permitted to wear special robes and personally received by the Kaiser. It is bizarre that Einstein and Fritz Haber, both Jews, were offered the first directorships of the two institutes respectively. To lure Einstein to Berlin, Planck and Nernst travelled to Zurich to spell out the terms of his appointment: he would be freed of any teaching burdens, provided with a liberal salary, and live in the fashionable Dahlem section of Berlin, not far from Lake Wannsee. To sweeten the inducement further, Einstein would become a member of the Prussian Academy of Sciences, granting him automatic citizenship. Einstein accepted these terms, provided he could concurrently keep his Swiss nationality. Thus Einstein entered into the Berlin mainstream of physics and, as expected, the break came with his first wife Mileva. One of the terms of the divorce ceded to her any future Nobel Prize money.

In 1914 an expedition headed to the Crimea to film the eclipse and try to corroborate Einstein's gravitational prediction. Unfortunately the war started and the Russians seized the equipment. Fritz Haber responded to the patriotic fervor and helped the war effort by producing artificial gasoline and devising a new method to make ammonia, the latter to earn him a Nobel Prize in 1918. Einstein remained unmoved, refusing to sign a war manifesto supported by Planck. Instead he urged international understanding, a correct action but out of step.

In the midst of the Great War, a friend of Einstein's, Felix Adler, became incensed with politics and assassinated the Austrian Prime Minister. Einstein provided a character reference at the trial and continued to correspond with Adler after he was sentenced to imprisonment. The scientist remained loyal to his friends.

In 1917 Einstein produced a paper on stimulated emission that took 43 years to impact on society. When an atom is in a excited state, a photon of the same energy can catalyze the atom to return to its ground state, emitting a photon that is a clone of the stimulating photon as to phase, direction and frequency. This became the basis for the laser in 1960.

At the end of the war he and Max Born tried to quell a pacifist riot. Mostly students, the rioters responded to Einstein's plea for order since he had a good anti-war record. Einstein and Born entered the Reichstag to plead the student's cause but it was futile - two naïve scientists in an emotional response to disruly students hardly merited serious attention.

The year 1919 was a turning point in Einstein's life. He married his widowed cousin Elsa, and Eddington reported the first measurement of the gravitational attraction of starlight by the sun at a joint meeting of the Royal Society and Royal Astronomical Society chaired by J.J. Thomson. Overnight Einstein achieved world fame. But if ever a man was ill equipped for the notoriety that befell him, Einstein is the classic example. His looks, his innocence and his inexperience in the world of press and politics spelled disaster. His suddenly enhanced reputation and pacifist views only served to irk the nation's industrialists.

The considerable mail he received he hung on a hook, and burned when the hook was full. He couldn't respond to it all and, fortunately, Elsa took over this chore. In great demand for every conceivable charity, he couldn't say no and Elsa had to intercede. To raise money for the starving children of Vienna, she began to charge a fee for his interviews and autographs. There were large queues to hear him lecture even though his theories were incomprehensible to the public. There were also disappointments, as when Eddington invited him to receive a gold medal from the Royal Astronomical Society but was forced to withdraw the invitation because of postwar anti-German feeling. Einstein was the only German invited to the 1921 Solvay Congress but he declined since he could not condone such discrimination.

On the lighter side, his popularity produced its share of humor:

> There was a young Lady named Bright
> Whose speed was much faster than light.
> She went out one day
> In a relative way
> And came back the previous night.

The *Scientific American* announced a $5000 prize for the best essay explaining relativity. Einstein reported that he was the only one in his circle not entering. Children were being christened Albert in alarming proportions and an Einstein cigar appeared on the market (presumably superior to the one Laue dropped into the river). In one of the popular English magazines, a cartoon depicted a bobby catching a thief with a flashlight that bent its beam around the corner.

At a restaurant at which he was not known, Einstein asked the waiter to read him the menu since he had forgotten his glasses. The waiter responded that he couldn't read either. As part of a string quartet Einstein was accused of not being able to count. In England the famous lines of Pope written as an epitaph for Newton:

Nature and Nature's law lay hid at night.
God said, 'Let Newton be,' and all was light.'

were completed with the lines:

It did not last: The Devil howling, 'Ho!
Let Einstein be!' restored the status quo.

Even Lord Rayleigh had his fun, suggesting that relativity proved the Norman invasion had not yet occurred.

When he found some free time, he began work on a theory that would unite electricity and gravitation. To illustrate, the mass of an electron can be equated to its electric charge, yielding a size for the electron. But how does gravity fit into this picture? His ideas about curved space defied simple schoolboy explanations, although he would speak of a blind beetle walking on the surface of an orange and thinking it flat. It was all too overwhelming for the masses but nonetheless they believed it as if the Messiah had arrived. Following the horrors of war, the world was looking for a lily-white postwar hero and found one in Einstein. Although seeking seclusion to develop his unified theory and to sail, he was drawn into the cause of Zionism. He became a world traveller unable to resist pleas for his help.

The first stings of anti-Semitism appeared in a campaign led by Nobel Laureates Stark and Lenard. Ironically, in a mood of forgiveness and to please Planck, Einstein reaffirmed his German citizenship. He even attended an anti-relativity lecture and sarcastically applauded the speaker. He was almost blind to the intensity of the incipient feelings growing around the country.

A visit to postwar France presented a ticklish problem since its loss of 1,350,000 men affected every family and anti-German feeling abounded. Langevin accompanied Einstein on the train journey into Paris. To avoid a hostile crowd at the station the two disembarked on the wrong side of the platform and escaped into the Metro. This cat and mouse game delighted Einstein. The French press damned him, and for security reasons only ticket holders were permitted into the lecture hall. Oddly enough it hadn't escaped the Parisian mentality that the Dreyfus Affaire should have been a constant warning against anti-Semitism.

In England, Einstein became disturbed over the anti-German *Lusitania* medals struck to protest the U-Boat sinking of the Cunard passenger liner with the loss of 1200 lives. The ship was actually carrying explosives for the English, contrary to Geneva Convention regulations. While always motivated by the noblest of intentions, no one took him very seriously in non-scientific matters. He was a good drawing card and made good copy, sometimes just because he was misunderstood when not speaking German.

While Einstein was visiting America in 1921-1922 Rathenau, a leading Jewish politician, was assassinated in Germany and fears were raised about Einstein. Indeed it was rumored that there was a price on his head. As long as he remained in the limelight the rabble in an unsettled country was a menace. Einstein was sufficiently disturbed to threaten to leave the Kaiser Wilhelm Institute but yielded once more to Planck's plea to remain loyal.

To add to his notoriety he was awarded the Nobel Prize in 1921, not for the controversial theory of relativity, but for the photoelectric effect. A comic drama ensued as the Swiss and German Ambassadors in Stockholm fought over protocol. Was Einstein Swiss or German? A compromise was reached: the Swiss Ambassador was to be present in Stockholm for the award and the German Ambassador would present the check in Berlin. Einstein did not hold the money long - it all went to Milerva as part of the divorce agreement!

However this was offset by income from speaking engagements. Wherever he went he was lionized by society as he became a showpiece for any hostess who captured him. The disrumpled genius referred to these occasions as 'feeding time at the zoo.' At a formal dinner party arranged by Eddington one guest provided some doggerel:

'The Einstein and the Eddington
Were counting up the score;
The Einstein card showed ninety eight
And Eddington's was more,
And both lay bunkered in the trap
And both stood up and swore.

I hate to see, the Einstein said,
Such quantities of sand;
Just why they placed a bunker here
I cannot understand;
If one could smooth this landscape out,
I think it would be grand.

If seven maids with seven mops
Would sweep the fairway clean
I'm sure that I could make this hole
In less than seventeen
I doubt it, said the Eddington,

Your slice is pretty mean.

* * *

The time has come, said Eddington,
To talk of many things;
Of cubes and clocks and meter-sticks,
And why a pendulum swings,
And how far space is out of plumb,
And whether time has wings.

I learned at school the apple's fall
To gravity was due,
But now you tell me that the cause
Is merely G mu nu.
I cannot bring myself to think
That this is really true.

* * *

And space it has dimensions four,
Instead of only three,
The square on the hypotenuse
Ain't what it used to be.
It grieves me sore, the things you've done
To plane geometry.

You hold that time is badly warped,
That even light is bent;
I think I get the idea there,
If this is what you meant;
The mail the postman brings today,
Tomorrow will be sent.
The shortest line, Einstein replied,
Is not the one that's straight;
It curves around upon itself,
Much like a figure eight,
And if you go too rapidly
You will arrive too late.

But Easter Day is Christmas time
And far away in near,
And two and two is more than four
And over there is here,

You may be right, said Eddington,
It seems a trifle queer.'

His return to Germany met with a bad press. Einstein, ever supported by a sense of humor, said, 'I am a foul-smelling flower but the Germans still wear me in their buttonholes.' During the next few years quantum mechanics exploded on the scene and left Einstein in a minority amongst theoreticians in opposition to its concepts. At the 1927 Solvay Congress he and Bohr had their first tussle over the uncertainty principle. Furthermore, a theory that relied on a probability description disturbed him. His famous line that God does not play dice is often quoted.

The strain of his commitments, now including the League of Nations, took its toll and he became aware of a heart problem in 1928. Elsa tried to control his movements but found it impossible. His pacifist utterances were widely quoted. In a 1930 speech at the Ritz-Carlton in New York he pointed out that if 2 per cent of military conscripts refused to serve, all armies would fail to exist. Jailing this 2 per cent would strain the military to a point of ineffectiveness. Short of this, he recommended alternative service for conscientious objectors.

Einstein took on the organized churches by defining religion as searching for an exit where no door existed. Asked whether $E=mc^2$ was consistent with religious teaching, he said there must be something behind the energy. Needless to say, his remarks on pacifism and religion made enemies everywhere.

In Germany a comic opera unfolded. The Berlin authorities decided to honor him with the gift of a house they had selected. After presenting Einstein with the deed in a well-reported ceremony, it was discovered that the current tenants had a long term lease! He was then offered a piece of land, only to later discover an ordinance forbade building on the property! In a third attempt they offered to purchase any piece of land selected by Einstein. He accepted, but when it came to a vote of the appropriate city council they turned it down! Einstein politely declined any further efforts on his behalf and bought his own piece of land, having earned enough from speaking engagements to do so.

At this time the anti-relativity clique in Germany, spurred on by Lenard and Stark, published a book 100 Authors Against Einstein. Einstein informed the press that if relativity was wrong they would only need one author. On a trip to Holland he was unexpectedly invited to meet the Belgian Royal family. Quickly borrowing appropriate formal attire, the two Alberts, King and scientist, met and began a life-long friendship.

Painters and sculptors were incessantly after him. Asked by a stranger what he did, he said he was a model. He was pursued for endorsements for such products as disinfectants, musical instruments and clothes! For a man often attired in baggy pants and no socks one wonders about the clothes. Elsa now charged one dollar for autographs, the money to go to charity.

Einstein attended the 1930 Solvay Congress where the participants arrived expecting to see a boxing rematch between him and Bohr. Einstein came prepared and he sketched on the board an experiment meant to defy the uncertainty principle. A photon, inside a box that is suspended by a weighing balance, is permitted to escape through a shutter at a precise time. The weight of the box after the event provides the energy of the photon from its effective mass. *Viola!* Energy and time are determined simultaneously. All eyes turned to a speechless Bohr. He had no answer and spent a sleepless night. The next day he showed that Einstein had forgotten to include the uncertainty in the clock due to the relativistic effects of the balance moving in the gravitational field! Most of the audience probably found the arguments too ponderous but they all enjoyed witnessing a good fight. The dispute continued during their lifetimes.

With the rise of Hitler, world events became a backdrop to the story of relativity and its principal actor. He was courted by the English and Americans, accepting an offer from a philanthropist to join a new Institute for Advanced Study at Princeton. When Hitler took office in January 1933 Einstein held a press conference in Los Angeles to announce that he would not return to Germany. (As soon as he finished speaking the earth shook in a severe tremor.) He resigned his appointment in the Prussian Academy of Sciences and even Planck was forced to issue a statement commending this as an honorable action to save embarrassment for Einstein's friends. Germany confiscated Einstein's property, including a sizeable bank account, and renamed Einsteinstrasse in Ulm, his birthplace. In May there was a massive book burning of the works of Thomas Mann, Stephen Zweig, and Einstein. Fritz Haber, hero of the Great War, Director of the Kaiser Wilhelm Institute for Chemistry, Nobel Laureate, 'blond Aryan' and baptized, was sacked and immediately offered a job in Palestine. He died in Switzerland en route. (The Berlin Institute has since been renamed after Haber.)

In one last trip to Europe, Einstein stopped in Belgium where he formally yielded up his passport and citizenship at the German Embassy, retaining his Swiss allegiance. He now stated to authorities at border crossing points that he had no fixed address. In all the years the Nazis spewed forth Aryan supremacy, Einstein was the one thorn in their side. Except for the efforts of the unimportant older physicists Lenard and Stark, it was impossible to discredit relativity. Heisenberg and others could not abandon physical reality; they taught the subject but didn't mention Einstein.

Einstein did an about face. Seeing Hitler as a menace that justified the use of force, he not only renounced pacifism but all links to communism, real or imagined. After accepting the Princeton offer the American Women's League pressured the State Department to deny Einstein a visa, claiming he was an atheist and a communist. That might have worked during the McCarthey era of the 1950's - it didn't in 1933.

His physics was still news and it was reported in the press that he was to publish a unified field theory. The leading newspapers were alerted and the *New York Times* hired a group of physicists to decipher the equations as they came over the wire. It made the news but had a short lifetime as errors were soon discovered. He kept trying but never succeeded! He left Europe in the autumn of 1933 for the last time, stopping in England to deliver a lecture. An assassination rumor packed the hall (actually a hoax to fill the house) and, as a final accommodation to posterity, he sat for the sculptor Epstein. This inspired the limerick:

> Three wonderful people called Stein
> There's Gert and there's Ep and there's Ein.
> Gert writes blank verse
> Ep's sculptures are worse
> Any nobody understands Ein.

At Princeton, and with Elsa's encouragement, he isolated himself from politics and scientific meetings. Laue visited him and was offered the opportunity to remain in America. He hated the Nazis but felt he had to return. Elsa died in 1937 and her administrative duties were taken over by Einstein's secretary.

The role Einstein played in bringing the threat of nuclear fission to Roosevelt's attention is well known. It accelerated the time scale by at least six months. No doubt the war would have otherwise ended before the tragedy of Hiroshima. Later, Einstein reflected on this incident as the one act in life he regretted. The letter to Roosevelt, composed by Szilard and Wigner and signed by Einstein, is worth reading:

Sir: Some recent work by E. Fermi and L. Szilard, which has been communicated to me in manuscript, leads me to expect that the element uranium may be turned into a new and important source of energy in the immediate future. Certain aspects of the situation seem to call for watchfulness and, if necessary, quick action on the part of the administration. I believe, therefore, that it is my duty to bring to your attention the following facts and recommendations.

In the course of the last four months it has been made probable, through the work of Joliot-Curie in France as well as Fermi and Szilard in America, that it may become possible to set up nuclear chain reactions in a large mass of uranium, by which vast amounts of power and large quantities of new radium-like elements would be generated. Now it appears almost certain that this could be achieved in the immediate future.

This new phenomenon would also lead to the construction of bombs, and it is conceivable - though much less certain - that extremely powerful bombs of a new type may thus be constructed. A single bomb of this type, carried by boat or exploded in a port, might very well destroy the whole port together with some of the surrounding territory. However, such bombs might very well prove to be too heavy for transportation by air.

The United States has only very poor ores of uranium in moderate quantities. There is some good ore in Canada and the former Czechoslovakia, while the most important source of uranium is the Belgian Congo.

In view of this situation you may think it desirable to have some permanent contact maintained between the administration and the group of physicists working on a chain reaction in America. One possible way of achieving this might be for you to entrust with this task a person who has your confidence and who could perhaps serve in an unofficial capacity. His task might comprise the following:

(a) To approach government departments, keep them informed of further developments, and put forward recommendations for government action, giving particular attention to the problem of securing a supply of uranium for the United States.

(b) To speed up the experimental work which is at the present being carried on within the limits of the budgets of university laboratories, by providing funds, if such funds be required, through his contacts with private persons who are willing to make contributions for this cause, and perhaps also by obtaining the cooperation of industrial laboratories which have the necessary equipment.

I understand that Germany has actually stopped the sale of uranium from the Czechoslovakian mines which she has taken over. That she should have taken such early action might perhaps be understood on the ground that the son of the German Undersecretary of State, von Weizsäcker, is attached to the Kaiser Wilhelm Institute in Berlin, where some of the American work on uranium is being repeated.

> Yours very truly,
> A. Einstein

The son referred to, Carl Freidrich von Weizsäcker, was a theoretician and student of Heisenberg, who was interned at the end of the war with nine other German scientists working on atomic energy. It is ironic that the first director of the Kaiser Wilhelm Institute, A. Einstein, should have been partly responsible for accelerating the American uranium effort now being independently continued at the very same institute. Furthermore, when the Allied bombing of Berlin forced Heisenberg to move the group to safety in southern Germany, they chose the very town where Elsa was born - Hechingen.

As the time approached to drop the bomb in 1945, Einstein wrote another letter to the President urging restraint. He advised that the bomb would provide America with an advantage over the Russians for many years to come. If the bomb had not been tested or employed, could it have been kept secret? It appears unlikely, since postwar congressional investigations of the huge expenditure of funds would have reached the press. Until his death in 1955 Einstein preached disarmament and control of nuclear energy. While he disagreed about the uncertainty principal, in the matter of atomic energy regulation he was in total accord with Bohr!

In the field of optics Einstein achieved three major breakthroughs. His light-quantum hypothesis accounted for the photoelectric effect and demonstrated that under certain conditions light behaved like particles with energy h times their frequency. He discovered stimulated emission, the means by which photons clone themselves in a laser, his relativity theory explained the zero aether drift - light did not need a conveying mechanism as sound does.

Between his letter to Roosevelt that accelerated bomb development, his discovery of stimulated emission that led to laser weapons, and his support of Zionism that partly contributed to the mid-East crisis, this benign pacifist and supporter of disarmament became a prank of the devil! Einstein, himself, remarked that everything he touched turned into a circus. Everyone wanted to hear him talk about relativity - he only wanted to play the violin for them.

Baggy pants, no socks, pockets full of child-like odds and ends, tobacco ashes everywhere, an uncontrolled bushy mane of hair, a sailor, a bicyclist but not a driver of cars, he, Maxwell, and Newton examined the inner recesses of their mind and established a pipeline to the Deity. On Einstein's death a cartoon appeared depicting the earth viewed from space with a single billboard message:

EINSTEIN LIVED HERE

The photoelectric effect, discovered in 1887, has stymied physicists who were leaning toward the wave theory of light. How a broad spread-out wave could have its total energy absorbed by a single wee electron was difficult to fathom. Physicists joked that on some occasions light behaved like a wave, on others like a particle:

On Monday, Wednesday, and Friday
Waves boasted, "It's my day,"
On Tuesday, Thursday, and Saturday
Particles sang, "Do it my way."
Sundays, though, were the very best
The good Lord said, "It's a day of rest."

The photoelectric effect was fairly simple. If visible light struck a metal it was generally reflected without change in color but if light at the higher energy violet end of the spectrum impinged on the metal an electron was ejected whose energy equalled the energy of the photon minus a threshold energy i.e. the energy needed to pull the electron away from its metallic bond. Every metal had its own threshold energy, sometimes called the work function. The technique for measuring the energy of the ejected electron was first developed in 1913 and it later became an important branch of physics called photoelectron spectroscopy. Yet this same beam of visible light could be diffracted by a grating or by slits. Photons and electrons confused man as much as their own children. It took a few years before physicists stopped asking the question 'wave or particle'? and permitted the photon to behave as the Creator planned.

But if one insists on a picture, imagine yourself an electron in a metal. Along comes a wavelike photon spread out over the entire countryside - you rope it and pull it in. If the photon has too much energy and you hang on, out you go flying off into space. Well, you asked for it.

During the first two decades of the twentieth century Niels Bohr came to the fore, aided by Sommerfield in Munich with an explanation for quantized energy levels by invoking a picture of electrons confining themselves to fixed orbits around the nucleus - a bit like planets revolving around the sun. These electrons could be switched from one orbit to another by absorbing or emitting photons of the exact energy difference. Certain simple rules were cooked up that explained the observed data but it was not until the third decade of the century that the quantum mechanics of Schrödinger and Heisenberg placed the field on a sound mathematical basis. Yet, as simple as Bohr's theory appeared to be it earned him the Nobel Prize in 1922.

The first two decades of the twentieth century are distinguished by the introduction of quantization and relativity, the third by quantum mechanics.

CHAPTER IX

Tutorial 3
Riding A Photon

We shall resort to a good dollop of literary license in order to take a slow ride on a photon - we don't wish the reader to increase his mass without the benefit of gastronomy. Besides, the author prefers a slow pace since he doesn't type very rapidly.

We shall select photons at three energies, 1. radio frequency, 2. visible, and 3. X-ray, to illustrate the phenomena that occur when real photons impinge on three types of materials, A. a metal, B. a ceramic, and C. a polymer. These nine possibilities account for the bulk of the situations encountered in our everyday world. Let us call these three photons Dean Alpha, Prof. Beta, and Dr. Gamma to make it sound scientific. Dean Alpha, the radio wave, has a wavelength of about 6 meters and an energy less than a millionth of an eV. Prof. Beta, the visible photon, has a wavelength of about 600 nm and an energy of about 1.5 eV, while Dr. Gamma is an X-ray with a wavelength of 0.1 nm and an energy of 10KeV.

The metal will be a thin foil of copper, the ceramic a plate of glass, and the polymer a sheet of polyethylene. We start our journey with Dr. Gamma heading toward the copper foil. All aboard!

Of all the electrons in their quantized orbits on copper the one most strongly bound to the nucleus has an energy of less than 9000 eV so that Dr. Gamma is strong enough to displace any of the electrons. If, indeed, one of the strongly bound electrons becomes the victim, it will find itself propelled outward with about 1 KeV, i.e. the energy of Dr. Gamma minus the binding energy of the electron (9 KeV). With all his energy gone Dr. Gamma disappears without a trace (habeas corpus notwithstanding). There is now a hole in this copper atom and a series of quantized jumps take place as the remaining electrons attempt to restore the atom to its undisturbed state. With each jump the electron gains energy and releases it by sending out a photon. Before too long the atom looks like new, the healing process is perfect. The 1 KeV energy of the original electron displaced by Dr. Gamma has been converted into heating the atoms in the neighborhood while the remaining 9 KeV energy has been emitted as photons. This overall process is called absorption.

Dr. Gamma could have suffered other fates some of which would enable him to survive his encounter with the piece of copper. The atoms of copper are arranged in a cubic crystalline form such that a copper atom is at each of the eight corners of the cube and a copper atom is at the center of each of its six faces. This cubic structure extends in all three directions. By virtue of Dr. Gamma's wavelength of 0.1 nm, which is comparable to the distance between copper atoms, a diffraction pattern develops with Dr. Gamma scattered without energy loss through any one of a set of specific angles. By directing a large number of Dr. Gammas at the copper foil one can record the set of specific angles at which they are scattered and from it deduce the arrangement of the atoms, i.e. the crystal structure of the copper. This process is called X-ray diffraction and has enabled scientists to determine the structures of zillions of substances. The International Union of Crystallography oversees world-wide interest in the subject.

The third adventure for Dr. Gamma is the Compton Effect, a billiard ball collision with one of the electrons. In this case Dr. Gamma will lose a few percent of its energy to the electron it hits but will otherwise recoil intact. Just before the collision the electron will have been moving in a specific direction and at a specific velocity. But the electrons in the piece of copper possess a range of velocities and directions, described by quantum mechanics. These can be deduced by measuring the energies of the recoiling Dr. Gammas. We can then compare them to the theoretical calculations deduced from the Schrödinger equation and see if our model of the copper atom in its metallic form is as we envision.

Between absorption, diffraction, and the Compton scattering of X rays a detailed picture of the electron and atomic arrangement in copper is possible. To those interested in the basic physics of any element Dr. Gamma is a valuable tool. But that isn't all. We can also do radiography. Suppose the piece of copper has a void inside that cannot be seen by external visual inspection. By measuring the absorption with a fine beam of Dr. Gammas the portion of the copper containing the void will show smaller absorption since the void is devoid of electrons and Dr. Gamma sails right through. Likewise the presence of an impurity like lead will have higher absorption since lead has more electrons per atom than copper. A photographic plate behind the copper will reveal details of any such imperfections.

One can also perform chemical analysis. If the copper sample contains other elements like nickel the absorption process will activate an X-ray whose quantized energy differs from copper. Each element has its own X-ray signature and this process, known as fluorescence, has become standard for chemical analysis. It is attractive because the sample need not be destroyed as in wet chemistry.

There is yet another series of measurements we can perform, subtle and interesting. Ordinarily the index of refraction of a material gives the ratio of the speed of light in a vacuum to the speed in the material as well as a measure of the angle through which it is bent on passing through an interface of air and the material. Window glass has an index of refraction of about 1.5 for sunlight, indicating that light is slowed down by 50% in passing through the window and that a light beam bends into the glass at the interface. A picture of the process is that there is a time delay as the photons pass the electrons, hence it takes longer for the photons to make a journey through glass.

If we consider Dr. Gamma traversing copper we find that the index of refraction is less than one by about one part in ten thousand. This does not mean that light travels faster in copper than in air, it always travels slower in materials than in a vacuum, but it does mean that the angle of bend is less as it enters copper from the air and that at angles less than a few minutes the copper acts as a mirror by reflecting Dr. Gamma (angle of incidence equals the angle of reflection). There are time delays as the photons pass the electrons in copper but the phase of the wave differs for different quantized electron orbits and it only appears as though the light has increased its speed. It may sound a bit bizarre and is probably why the subject is avoided in textbooks but do think about it. You may even begin to feel you understand it but if you don't, find a physicist with a new Ph.D. and ask him to explain it without equations. You may enjoy the encounter.

If the copper foil is replaced by the glass Dr. Gamma is able to undergo the same diffraction with but one difference - the glass is not crystalline and the arrangement of the atoms is not easily deduced. Only an approximate solution is possible. If we examine the polymer polyethylene Dr. Gamma will find it somewhat similar to copper in that it appears crystalline, although not the same in all three directions. Polymers are typical of natural products like wood, hair, leaves, bamboo, celery, etc. in that there is a long chain of strongly bound atoms in one direction while the bonds are weaker in the two other directions. This property serves trees well in enabling them to achieve great height and bend in a gale. It also accounts for the crunch when you bite into celery. We now give Dr. Gamma a rest.

Prof. Beta is a nice yellow photon, not too energetic yet bright and visible. As we approach the copper foil we run smack into a wall. There is a bunch of electrons waiting to gobble up Prof. Beta and spit him out again. These are the very electrons that make copper behave like a metal and conduct electricity. They are free to roam about the entire piece of copper and are called the 'conduction band', like a gang of hoodlums guarding their territory. They occupy a set of quantized states very close together. Almost every time Prof. Beta approaches the surface of the copper he is absorbed and re-emitted like a mirror.

Ah, but exchange the copper for a piece of glass and Prof. Beta goes sailing through, stopping briefly to greet each electron, and proceeding on the same path along which he entered. Each electron is so rigidly attached to its atom that Prof. Beta's 1.5 eV energy can do no more than tickle each electron. Stopping to shake hands or give a friendly pat takes time and this slows the Prof. down by 50%. What a difference from the foil of copper!

Approaching the polymer polyethylene Prof. Beta is greeted just as with glass but not as long so that he's slowed down half as much. Otherwise, things are the same as in glass. The electrons are too strongly bound for Prof. Beta to displace any of them.

Switching to Dean Alpha, the radio wave, and donning ear muffs in case there's some loud rock and roll blasting forth, we line up the piece of copper again. The same old treatment as for Prof. Beta - a slap in the face and Dean Alpha is sent reeling back by the band of conduction electrons. Switch to the glass or the polymer and Dean Alpha is again treated like Prof. Beta - he passes through with but a slight pause.

Having tried all nine options, it's now time for a break, have a cup of tea, and contemplate an interesting query. Why is copper reddish in color, gold and brass yellow, and most other metals silvery? If Prof. Beta had been more energetic when he approached the copper foil, say 2.2 eV, he would have suffered a different fate. He'd have been grabbed by a conduction electron and made to disappear! Poof! In fact all photons of energy greater than 2.2 eV would have been subjected to this same treatment but not those of lesser energies. Hence, when white light strikes the copper the yellow, green, and blue photons are absorbed while the reds are reflected. With gold and brass the energy for absorption is shifted slightly higher than 2.2 eV so that yellow and red are reflected, green and blue absorbed. In the other metals all visible wavelengths are reflected and they appear silvery or whitish.

This explains why substances are colored. If a substance is red all visible photons are absorbed raising the electrons to a higher energy quantized state. When the electrons return to the quantized ground state the energy is released as a red photon. Different substances have different quantized energy levels leading to different colors. Weavers, artists, etc., have spent lifetimes searching for natural and artificial dyes to delight the eye never dreaming that quantized energy levels were the key to their behavior.

CHAPTER X

Early 20th Century
Part 4

Niels Bohr was showing a scientist around his garden when the visitor noticed a horseshoe nailed to the door of the tool shed. 'Surely you don't believe in these things?' enquired the guest. Bohr responded, 'The nice thing about horseshoes is that they bring luck even if you don't believe in them.'

A few years ago at a small conference in London marking the centenary of the birth of the famous Nobel Laureate, a number of prominent physicists spoke about the great Dane and his idiosyncrasies that have attained legendary proportions. Just before the meeting adjourned, someone arose to say that, in listening to all the stories about Niels Bohr, he did not recognize the man. He identified himself as Erik Bohr, engineer and Niels's son.

Such is the legendary fame of this stocky Dane with bulbous head who was responsible for much of the exciting developments in modern physics, and whose life was replete with drama and adventure. No one who crossed his path fails to have a favorite story.

Mine occurred during a lecture in 1952 when I listened to the famous mumbler (he was difficult to understand). During his lecture he referred to the magnetic moment of the electron and extricated a small two-inch square of paper from his waistcoat, copied the formula $eh/2mc$ on to the blackboard, and returned the paper to his pocket. This simple expression is universally known as the Bohr Magneton! I enquired of a Danish physicist whether the great man was more coherent in his native language, but was assured he also mumbled in Danish. The ever-present pipe didn't help.

Born on 7 October 1885, his father was Professor of Physiology at the University of Copenhagen while his mother came from a prominent Jewish banking family. Niels showed outstanding skill with his hands and turned into a good sportsman and skier. He studied physics and graduated from the University of Copenhagen at 22 with a specific interest in surface tension.

J.J. Thomson, Cavendish Professor at Cambridge, had discovered the electron in 1887 (the toast at the annual Cavendish dinner expressed the hope that no one would find a use for the little beast)! Bohr's doctoral thesis on the electron theory of metals was granted in 1911 and he decided, with private funding, to go to Cambridge and study with Thomson. At his first interview the young and confident Bohr did not hesitate in pointing out certain errors in one of Professor Thomson's papers, occasioning a lasting impression on the English Nobel Laureate and quickly making the rounds as after-dinner gossip. The Dane benefited considerably by interacting with the famous scientists who visited Cambridge, particularly Ernest Rutherford. Bohr would never forget the parody he learned about Thomson and the Cavendish (tune of Clementine):

> In the dusty laboratory
> Mid the coils and wax and twine,
> There the atoms in their glory,
> Ionize and recombine.

> CHORUS

> Oh my darling, oh my darling,
> Oh my darling ions mine.
> You are gone and lost forever
> When just once you recombine,

Rutherford, a New Zealander, was sufficiently impressed with Bohr to invite the Dane to Manchester in 1912 to join in the experiments on the atomic nucleus, discovered by scattering alpha particles from atoms. On his return to Copenhagen Bohr married and received an appointment as an assistant in physics.

Unravelling the structures of the atom occupied the attention of many laboratories and centered on the discrete optical lines emitted by the simplest atom hydrogen. Classical pictures of planetary motion and electromagnetic theory would have had the electron constantly losing energy and being sucked into the nucleus. Bohr's model of quantized orbits was a good start towards an explanation but failed to convince classical theorists.

The first test for Bohr's new ideas came in 1913 at a meeting of the British Association for the Advancement of Science in Birmingham. As expected, the traditionalists objected with comments like: 'If we must use quantized orbits we might as well have no explanation at all!' Furthermore, Bohr's theory did not account for all the spectral lines. In September 1914 Bohr made a safe wartime passage to Manchester under favorable foggy and stormy weather conditions. Scientists were not deeply involved in the Great War, although Bohr was emotionally opposed to the Germans who had formerly annexed Schleswig Holstein from the Danes. Bohr arrived in Manchester to discover that Rutherford was stranded in Australia and he volunteered for extra duties in his absence. Rutherford made a safe journey back and Bohr enjoyed a successful two years before returning to Copenhagen in 1916, again under wartime conditions. Neutral Denmark continued its mail service with Germany and Bohr received a paper from the great Sommerfield at Munich expanding the work on quantized orbits and accounting for more of the optical spectrum of hydrogen.

Copenhagen under Bohr, like Göttingen under Max Born and Cambridge under Rutherford, became a focal point for the study of physics, attracting the well-known. In 1921 Bohr continued his work on the atom and was honored by the Royal Society. The following year he received the Nobel Prize. The work was still incomplete and would require contributions from Pauli, Goudsmit, Uhlenbeck, Schrödinger, Heisenberg and Dirac before all the optical levels of atoms were understood within the framework of quantum mechanics.

The overly taciturn Dirac came to Copenhagen for a year's study and Bohr described him by telling the story of a man who returned a parrot he bought because it did not speak. The shopkeeper countered that he had sold the customer a thinker.

Following the first years after 1926, when both Schrödinger and Heisenberg presented viable theories of quantum mechanics, Bohr was frequently asked which of the theories to employ. Bohr told the story of the boy who asked the shopkeeper for a pennyworth of mixed candy. The lad was given two pieces and advised to mix them himself.

Gamow, of Russian origin, knew both Rutherford and Bohr at Cambridge and produced this poem to relate something of their personalities as seen during a meeting in Cambridge:

> 'That handsome hearty British Lord
> We know as Ernest Rutherford
> New Zealand farmer's son by birth,
> He never lost the touch of earth;
> His booming voice and jolly roar
> Could penetrate the thickest door,
> But if to anger he inclined
> You should have heard him speak his mind
> In living language of the land
> That anyone could understand!
>
> One day George Gamow, as his guest
> By Rutherford was so addressed
> At tea in honour of Niels Bohr
> (Of whom you may have heard before)
> The men talked golf, and cricket too;
> The ladies gushed, as ladies do,
> About a blouse, a sash, a shawl -
> And Bohr grew weary of it all.
> "Gamow," he said, "I see below
> Your motorcycle. You will show
> Me how it works? Come on, let's run!
> This party isn't any fun."

So to the motorcycle Bohr,
With Gamow running after, tore.
Gamow explained the this and that
And Bohr, who on the saddle sat,
Took off to skim along the Backs,
A threat to humans, beasts, and hacks,
But though he started full and strong
He didn't sit it out for long.
No less than fifty yards ahead
He killed the engine dead.
And turning wildly as he slowed
Stopped traffic up and down Queen's Road.

While Gamow rushing to the fore,
Was doing what he could for Bohr.
Who should like Jove himself appear
But Rutherford. In Gamow's ear
He thundered: "Gamow! If once more
You give that buggy to Niels Bohr
To snarl up traffic with, or wreck,
I swear I'll break your bloody neck!" '

The well-known Einstein-Bohr debate over quantum mechanics started at the famous fifth Solvay Conference of 1927 when the subject of complementarity, closely associated with the uncertainty principal of Heisenberg, became a much debated issue. Bohr and Heisenberg believed that it was impossible to determine certain pairs of properties exactly, such as position and momentum or time and energy, since any attempts to measure one of the pair disturbed the other. For example, if one uses a photon to locate an electron, the photon will transfer momentum to the electron and the position determination will not be exact.

Einstein continually dreamed up schemes to overcome these objections but neither he nor Bohr ever changed their minds. Einstein insisted that a proper description of an atom should yield a complete picture, not one resorting to a probability for locating the electrons. This prompted the famous Einstein remark that God does not play dice. Bohr's rejoinder was that one must exercise caution in ascribing such behavior to the Deity.

The debate reappeared at the 1930 sixth Solvay Conference where the spectators were keen to witness the rematch. Einstein suggested another clever experiment that left Bohr stumped. It required a sleepness night before Bohr realized that Einstein has omitted his own theory of relativity in describing the measurement. After almost 60 years Bohr's approach is still the accepted one but, in this everchanging field, the advantage may someday shift back. Let's hope the opponents can be notified.

With the discovery in England of the neutron in 1932 by Chadwick, it became apparent that this uncharged particle was capable of easily penetrating the nucleus and producing artificial transmutations. These experiments began to occupy the group in Rome under Fermi, those in Paris under the Joliot-Curies, while in Copenhagen neutron absorption was measured in gold with Bohr's Nobel medal.

Following this upsurge of interest in artificial radioactivity, a national subscription was undertaken and the Danish people presented Bohr with an expensive radium source. When mixed with beryllium it produced fast neutrons, and when immersed in water the neutrons were slowed down making them more effective in transmuting elements. When Einstein was asked if 'splitting the atom' might be useful, he replied that it was like a blind man hunting ducks at night by firing bullets straight in the air in a country where no ducks had been seen before.

The founder of the Carlsberg Brewery bequeathed his residence as a House of Honor for the most distinguished Danish citizen, and in 1932 Bohr and his family moved in. It abutted the Brewery and a certain fraction of the profits still go to maintain the Institute. Two years later tragedy struck when the Bohrs lost a son in a boating accident. Bohr carried on; his simple philosophy was that no one was without grief.

With the rise of Hitler and the excesses of the Nazis, Bohr turned his attention to the refugees forced to flee Germany. Following the discovery of fission in 1938 by Hahn and Strassmann in Berlin, there was set in motion a series of events that centered around Bohr and thrust nuclear physics into the forefront of warfare and politics.

When Otto Hahn and Fritz Strassmann discovered barium as one of the products of neutron bombardment of uranium, it befell two Jewish refugees to explain the phenomenon as fission. Lise Meitner, a colleague of Hahn's and then working in Stockholm, invited her nephew Otto Frisch, working with Bohr in Copenhagen, to spend the 1938 Christmas holidays in Sweden. Their Yuletide explanation assumed that when the nucleus of uranium absorbed a neutron it necked in the middle like a water droplet, and the repulsive positive charges of the two halves repelled each other and produced fission. The simple calculation of this 200 Mev electrostatic repulsive energy agreed with the known mass differences of the fission fragments (using $E=mc^2$). Frisch immediately communicated this to Bohr, and caught him just prior to his boarding a boat to America for a sabbatical at Princeton.

Just before this, Fermi (whose Jewish wife had become subject to the new Italian segregation laws) visited Bohr in Copenhagen and was informed confidentially that he would receive the Nobel Prize. He might want to consider keeping the money out of Italy to avoid its currency restrictions. After the award was announced, Fermi and his family departed from Rome permanently, leaving all their possessions behind, and headed for America. When Bohr arrived in New York, Fermi and his wife were there to greet him. (Within four years of that time Fermi had produced the first chain reacting pile.) After Bohr delivered a talk about fission at a conference in Washington in January 1939, pandemonium broke loose as physicists rushed to their laboratories to confirm the finding. Bohr first recognized that only the rare U^{235} isotope was responsible for fission.

Around 15 March 1939 three events occurred that had far-reaching consequences. Hitler moved into Czechoslovakia, not only presaging war but delivering to him their uranium mines, *Time* magazine blurted out the awesome potential of a uranium weapon so that the entire world knew, and a private conference was held by Bohr, Szilard, Rosenfeld, Wheeler, Wigner and Teller at which they decided to warn Roosevelt (through Einstein) of the threat such a weapon held in Hitler's hands. Bohr returned to Copenhagen to publish a detailed theory of fission that he and Wheeler had worked out during his stay at Princeton. A few months later the war started in Europe.

On 8 April 1940 Bohr was in Oslo to deliver a paper, and he spent that evening as dinner guest of the King, leaving on the overnight train for Copenhagen. He arrived home to discover that the Nazis had invaded both Norway and Denmark, a necessary move for Hitler to ensure the uninterrupted flow of Swedish iron ore along the Norwegian coast and through the narrow Kategat separating Denmark from Sweden. (As a bonus from the Norwegian occupation, Hitler had captured the only heavy water plant in Europe.) By smuggling a telegram into Sweden, Bohr advised Frisch in Cambridge not to return to Copenhagen and added: *TELL MAUD RAY KENT.*

When Frisch showed this to Military Intelligence they concluded it was in code. *RAY* undoubtedly meant radiation and *KENT* must be the target area. *MAUD* required a little more imagination but one guess was it stood for Military Applications for Uranium Detonation. This dire message could only mean the Germans had already developed a bomb! A secret committee on fission weaponry was hastily formed in England and named *MAUD*. Years later, when Bohr escaped to England, he informed the British that Maud Ray was the name of their children's nanny when they lived in England, and he wanted to tell her they were safe following the Nazi invasion. She lived in Kent.

With the invasion of Holland and Belgium and the fall of France, Bohr felt isolated. To make matters worse the Gestapo set up their headquarters next to the Bohr Institute. Nevertheless, he continued his nuclear research even managing to get an accelerator built in spite of materials shortages. One word from Bohr was all that was needed for the Carlsberg Brewery and the Danish shipyards to 'misplace' the necessary steel. With Gestapo spies everywhere the game of cat and mouse became part of Bohr's daily existence. On one occasion, a man desperately seeking shelter as an escaped prisoner presented himself to Bohr for help. Bohr had been alerted by a friend at Gestapo headquarters and avoided the trap.

One of the remarkable cloak- and -dagger stories to emerge from this period concerns a visit Heisenberg made to Copenhagen in the autumn of 1941. Heisenberg had been placed in charge of the German uranium effort at the Kaiser Wilhelm Institute but had neither the stomach nor the resources to develop a bomb. Rather, he hoped that a meeting with his old teacher might induce Bohr to seek a secret agreement with Allied scientists not to work on a weapon. Realizing that Bohr would be under surveillance, Heisenberg met him on a remote back street but Bohr was hardly in a receptive frame of mind for talking to a German. The moment Heisenberg mentioned the uranium weapon Bohr assumed he was seeking his cooperation and was immediately overcome with wrath, terminating the meeting.

More than likely, Bohr sent a warning about Heisenberg to the Allies. Considering the emotional strain of these two giants, it is no wonder that accurate details of the meeting have never been obtained, even from the principals. If this dialogue had been more amicable, it would have opened an interesting line of communication during the war. While Heisenberg recognized Hitler as a madman, he played a dangerous game. If his group had achieved a chain reaction, and they did come close, Hitler could have strewn tons of radioactivity over London in the V2 rockets.

As German atomic bomb development accelerated in 1943 the Gestapo decided to take over the Bohr Institute. Along with thousands of other Danish Jews, Bohr was warned and decided to escape Denmark after first destroying any incriminating evidence about his help to refugees. Laue's and Franck's Nobel medals (left with Bohr for safekeeping) were dissolved in acid (he already donated his own to the Finnish war effort) and innocently left with other bottles of chemicals, and Professor and Mrs. Bohr were rowed to Sweden, the boys escaping later. A high altitude Mosquito fighter was sent from England and Bohr was flown to safety, although he had a narrow escape when he failed to use his oxygen mask properly. On arrival he learned for the first time of the Manhattan Project and soon afterwards left for America.

The Gestapo were livid at Bohr's escape and sent Heisenberg and a Gestapo agent to determine if the Danish Institute had been doing war work. If so, the Institute was to be blown up. Heisenberg assured the Gestapo that this was not the case and saved the buildings and personnel. Shortly afterwards the Danish underground planned to blow up the Institute, afraid it was to be used for war work by the Nazis. Bohr in America was informed and advised against it.

Bohr and son Aage, with code names Nicholas and James Baker, arrived in New York by military ship. Niels was personally accompanied on the train to Los Alamos by General Groves, head of the Manhattan Project. Disappointingly, Bohr could provide Groves with no information about the German effort, and the torture of listening to Bohr mumble for four days was the most agonizing thing that had ever occurred to Groves! At Los Alamos Bohr's familiar figure was recognized by all and he immediately became 'Uncle Nick'. His mission was to offer any advice or comments on the program. However, the development was too far advanced for Bohr to be very helpful.

Bohr now devoted all his attention to postwar concern over nuclear developments. He visited Roosevelt and struck a sympathetic chord. Bohr was flown to London to confer with a reluctant Churchill about his ideas on nuclear regulation, at a time when Churchill was strained to capacity trying to win the war and plan for D-Day. With a mumbling Bohr speaking about postwar irrelevancies, Churchill brought the meeting to a rapid end and refused ever again to see Bohr.

All of Bohr's politicking over control of atomic energy went for nought and, until he died in 1962, he pleaded a fruitless case. If he had been a better speaker he might have been successful. His early Nobel work on the optical levels of atoms and his debates with Einstein make exciting scientific reading. His postwar political crusade points up the necessity for scientists to communicate more effectively.

At the end of the war the bottle of dissolved gold was returned to Sweden and new medals were cast. His son Aage took over the Institute after Bohr's death.

The most profound optical revolution in man's history occurred for the most part in Hollywood, where man's mind is controlled by subjecting him to a 'pleasurable' transportation into a world he ordinarily does not experience. Every conceivable situation is relived for him - he fantasizes himself as part of a universe that evokes images of new, imaginative, and totally absorbing situations in which he can vicariously perform roles that make him a hero for the moment, a king for an hour, a rich man for an evening, or a killer for a second. The technique relies on persistence of vision to merge individual pictures projected $1/16$ sec apart, the artistry of the writer, the creative ability of the actors and directors, all brought to reality by the liberal flow of money. TV has added a new dimension to the cinema by invading our homes and permitting the viewer to instantly witness events taking place all over the earth and even on the moon or in space.

Cinema emerged in the first two decades of this century, sound was added at the end of the third decade, and color at the end of the fourth. Today there is no bodily function during our waking hours that occupies as much of our time, that affects as much of our creativity and learning as viewing cinema and TV. Where this 'monster' is leading us is as impossible to predict as the invention of the laser and the transistor.

Fox Talbot, Muybridge, and Edison are responsible for the scientific breakthroughs while D.W. Griffith paved the way for finding interesting and psychologically effective conversions of photographic moving images to life's experience. To the cinema Griffith was as important as Monet, Rembrandt, and Picasso were to painting. He has fallen into obscurity because of advances in cinematic technology but during the first twenty years of this century he scaled the Everest of his art.

The first motion picture was viewed in 1895 at a time when the technology concentrated on discernible rather than meaningful images. It was soon realized that its effectiveness depended on telling a story in its own way, not by imitating a stage production or a novel. The strategies of close-ups, long shots, cuts, etc., were early avenues of directional exploration. The unit of structure was recognized to be the shot rather than a scene composed of shots. Advances in acting technique came by developing film stars rather than employing actors who were inhibited by their stage craft.

Critical to this technology is persistence of vision. The employment of persistence of vision to fool the mind dates back to 1825 when Dr. John Paris created a two-sided rotating toy with a parrot on one side and an empty cage on the other. By spinning the toy the Parrot appeared to be inside the cage. Edison entered the scene at the turn of the century by concentrating his resources on developing a camera that used flexible film, by constructing a studio to shoot the scenes, and by building up a nation-wide organization protected by patents. The film and shutter movement was crucial, i.e. the shutter remained closed while the film transport mechanism advanced to the next frame, whereupon the shutter opened momentarily while film transport ceased. Each frame was separated by an equal amount of darkness.

Edwin S. Porter's *The Great Train Robbery* in 1903 demonstrated the imaginative use of cutting back and forth between two scenes, allowing the mind to integrate the timing sequences, i.e. time did not have to be linear in a film. The brain can store one sequence of events while another earlier sequence is being viewed and then integrate the two.

D.W. Griffith started his career in films in 1907 as an actor for Porter, who was then shooting scenes in the Bronx. By the next year he was assigned to directing films in order to meet the great demand from the nickelodeons. The public wanted action and drama, even melodrama, i.e. the distortion of events to emphasize the conflict. In the next five years Griffith made 150 films, enabling him to try new ideas almost weekly. One of his early discoveries was the half shot or close-up - a cinematic adaptation of Rembrandt's use of light and shadow to force the viewer to focus his attention. He moved to the long shot to relate the actor to the entire panorama of the story. He showed that one can cut back and forth without the viewer losing his ability to follow the action. The brain became an interesting device for collating and differentiating light, images, and time. Camera movement on a dolly created the sensation of viewer movement even though the actors remained stationary.

Griffith had to sell upper management on the viability of cross cuts, citing Dickens's use of the technique in his novels. At this point in his directing career the greater light output of the arc lamp was introduced, two carbon rods that ionized the air between them. Griffith successfully explored the effectiveness of lighting a scene to create a mood, and departed from the practice of having the actors over-emote, even spending time in rehearsal. His technique of sharp cutting between shots tightened the dramatic content of his stories; editing became as important as shooting.

Itching to be rid of interference with his ideas Griffith joined an independent company in 1914 and set out to produce his masterpiece *Birth of a Nation* based on an antebellum novel 'The Clansman'. With a cast of thousands and a script that existed only in his head Griffith spent the unheard sum of $125,000 and took half a year to shoot and edit a film that extolled the Ku Klux Klan. Although the film was not anti-black it presented Griffith's personal view that the mixture of races was undesirable. Following the history of a southern and a northern family Griffith took the viewer through the vast battle panoramas of the Civil War, more vivid than to those who had served in the conflict. The grand scale of his effort proved successful at the box office and Griffith took a giant leap with his next film *Intolerance*. Budgeted at $2,000,000 he continued to exploit the success of huge casts and vast panoramas, interleafing several stories such as the fall of Babylon, the Passion, the anti-Huguenot movement in France, and an American tale of injustice. While it was a vast film in concept and execution Griffith's success depended on his focus of attention at the plight of individuals. While Griffith's successors like Bergman and deMille underscored the effectiveness of the media in employing light manipulation and persistence of vision to control the human mind Griffith was the trailblazer. During the time film-making was exploding on the scene another optical revolution was taking place in Paris, the refining of radium and its use in curing cancer.

Tragic 20th century stories of success and suffering in science call to mind Stephen Hawking, J. Robert Oppenheimer, and Madame Curie. Within the first two decades of this century the Polish-born Marie Sklowdowska overcame the enormous odds of male domination in French science and society to become the first person to win two Nobel Prizes. Her story and her contribution to optical technology are impressive.

The headlines of Le Journal for 4 November 1911 read: 'A STORY OF LOVE. MME. CURIE AND PROFESSOR LANGEVIN'. The Paris reporter went on to say, 'The fire of radium had lit a flame in the heart of a scientist, and the scientist's wife and children were now in tears.' The next day every Parisian newspaper related the scandalous tale of the lovers. The radiation exposure that had poisoned Marie Curie's body could not have struck as devastating a blow to the well-being of this famous scientist. 1911 was to prove the low point in her life.

For the next few years she withdrew into virtual isolation but her patriotism brought her back into the limelight. During the Great War it became imperative for doctors at military hospitals and emergency field stations to have reliable diagnoses of military wounds. In many cases the wounded were unconscious and unable to describe their injuries. Portable X-ray units housed in special trucks scurried up to the front lines to retrieve the wounded and perform rapid X-ray examinations. Driving one of these specially equipped vehicles was the petite, almost emaciated widow in black, Madame Curie.

She could easily have been mistaken for a Florence Nightingale but this female Nobel Laureate had not only been responsible for organizing the radiographic units in the French Army but also for training the technicians as well. A Lady of Mercy, she was an unselfish hardworking scientist. In a male-oriented society she had great obstacles to overcome, but her work ethic was so strong that she surmounted these difficulties as well as the Langevin scandal to become a world heroine. Yet, in spite of her achievements, the chauvinistic French Academy of Sciences, faced with a hostile press for considering a woman as a possible member, turned her down and her own pride did not allow her to stand a second time for election, where she probably would have gained entrance. Marie Curie was responsible for the earliest work on the high energy end of the electromagnetic spectrum. Unwittingly, she died from her own researches.

Marie Sklowdowska was born in Warsaw, Poland, in November 1867. Her father taught mathematics and physics and Marie developed into a bright studious pupil. Her mother died when she was 11 and this left the family in straitened circumstances. To make ends meet, Marie gave up her room to a boarder. She slept in the living room and had to tidy it up before anyone arose.

Marie won a gold meal for academic achievement in high school and also helped with the household finances by tutoring pupils. Working as a governess in her spare time, she supported her younger sister at school in Paris, the cultural center of Europe. Only after her sister and become an M.D. did Marie consider joining her. Using the last of her savings, Marie sat on a fourth-class wooden stool for the train journey across Germany. In Paris she took a written exam, and was granted a license to practice physics.

In 1895 she met and married Pierre Curie, a physicist interested in magnetism and later to give his name to the Curie point of a ferromagnet. Marie's first laboratory work was on the magnetic properties of cold-worked steel. She left the research briefly to give birth to their daughter Irène, born in 1897.

Returning to her husband's laboratory to work without pay, she became fascinated by the ionizing properties of the newest scientific curiosity, radioactivity. Pitchblende and thorium salts were employed but stronger sources were needed and she began her life's work in concentrating these materials. This led to the discovery of two new elements, polonium, named after her native Poland, and radium. The latter's long, essentially immeasurable, half life puzzled her since Rutherford had shown that all radioactivity decayed exponentially. However she had no equipment capable of measuring a half life of 1,690 years! Since its primary radiation was alpha particles, most of the energy was absorbed in the sample itself and this generated heat and light. The press reported that Madame Curie had discovered a source of perpetual motion since it did not appear to decay! Pierre estimated that a gram molecular weight of radium produced as much heat energy as a gram molecular weight of hydrogen combusted in air.

The Nobel Prize was awarded to Becquerel and the Curies in 1903 for discovering and researching radioactivity. Presaging the future, Pierre Curie warned in his acceptance speech that radium in the wrong hands could be used for sinister purposes. Yet it was still not recognized at that time that radiation poisoning was of biological concern. Having won the Nobel Prize and having worked six years without pay, Madame Curie was finally placed on the payroll just before the birth of their second daughter Eve.

A French industrialist set up a factory for the production of radium for medical use. The Curies refused to take out patents or accept any remuneration for their efforts; their intentions were nobly honorable.

About this time the brilliant physicist Paul Langevin entered Marie's life. Evidently the two became rather fond of each other. Whether this contributed in some way to the tragedy of 1906 we shall probably never know, but while crossing a street in Paris Pierre Curie was killed by a truck. Marie, undaunted in regard to her work, took over Pierre's lectures. She was an inexperienced speaker, and in her soft monotone she continued at the precise point that Pierre had left off in his notes. The novelty of a woman lecturer so intrigued the public that the galleries were packed with spectators but she remained unmoved by this crowd of gapers. Soon afterwards she was appointed to her husband's Chair, the first female professor at the Sorbonne.

Encouraged by this acceptance of her gender she made application to the stodgy French Academy. Like English Men's Clubs, this bastion of male domination was insurmountable and her application was turned down, with considerable support from the Fourth Estate. Perhaps partly in retribution, the 1911 Nobel Prize in chemistry was awarded to her for the discovery of the new elements polonium and radium, making her the first two-time winner. The money went to her research and her friends.

Following her front line activities in the Great War and with a world-wide reputation, a group of Americans subscribed to the purchase of a gram of radium. Having forgotten the scandal of 1911, a gala French send-off was arranged for her trip to New York. President Harding made the presentation to her in Washington and she brought the radium back to France to continue her research. On arrival in Paris only one person was there to great her at the station- everyone was home listening to the Dempsey-Carpentier fight!

But a life working with radium had its toll. She submitted to four operations for cataracts, and her hands were terribly distorted with radiation burns. Marie died in 1934, leaving her daughter to carry on the work.

Pierre was the more accomplished physicist with his discovery of piezoelectric crystals and his work on magnetism, but the world was captivated by the first woman competing with men in science.

What drove Marie? She tried to survive in a man's world and may have found shelter behind her test tubes. The laboratory, more so than her home, became her sanctuary. The one chore that combined the work of a mother and work in the laboratory was bathing her children. She refused to allow anyone to substitute for her.

1920 marks the end of two decades that prepared the scientific community for the revolution of the third decade, the de Broglie postulate, i.e. the electron had a wavelength and could be diffracted, the Compton effect and the discovery of quantum mechanics.

Turning skyward, spectral lines were recorded by astronomers and identified with substances known on earth. Helium gas was first isolated in 1895 and recognized as the principal gas on the sun. Stars were classified by their spectra as hot and blue or cold and red, suggesting a pattern relating to their age. Astronomers were puzzled at the source of a star's energy and it was Eddington in 1920 who made a profound comment based on Rutherford having split the atom in the Cavendish Laboratory, 'If indeed, nuclear energy in the stars is being freely used to maintain their great furnaces, it seems to bring a little nearer the fulfillment of our dream of controlling this latest power for the well-being of the human race, or for its suicide.' He died eight years before Hiroshima.

In 1908 and 1917 two huge telescopes were installed on top of Mt. Wilson in California in order to reduce interference from haze and atmospheric thermal effects. Working there Hubble discovered the famous Doppler red shift i.e. stars appeared to be moving away from the earth as if the universe were expanding. Twentieth-century astronomy was opening new vistas in our understanding of the scheme of things.

CHAPTER XI

Tutorial 4
Illusion and Reality

The 0.1 second time for persistence of vision (POV) has had a far-reaching impact on our lives. It is responsible for our brain smoothly blending the individual frames of a movie or a TV image into a lifelike scenario of action. Its employment is an integral part of the sports we watch or engage in. A pitcher throwing a baseball at 85 miles per hour reaches the batter in 0.5 sec, giving him little time to observe, evaluate, and react - only about 5 times the POV time. By causing the baseball to curve or drop in the last ten feet the batter has less than 0.1 sec to respond effectively. It may be that the better ball players have shorter POV or reaction times.

POV times also enter the age-old profession of prestidigitation. The shell game, three walnut shells and an uncooked pea, relies on both persistence of vision and on psychological factors as the expected outcome of certain movements of the hand and the pea as well as the magician quickly palming the small object when attention is focused on the rapidly moving shells.

The work of Harold Edgerton, inventor of the very high-speed stroboscope, provided the photographer with sufficient light to photograph a bullet in flight. By charging up condensers and discharging them rapidly into a light source motion is frozen. With the advent of pulsed lasers the time has been reduced to one millionth of the time of an Edgerton strobe i.e. to femtoseconds. These fast times enable us to freeze atomic processes, even high-speed chemical processes. No doubt if an Edgerton lamp were employed to photograph a magician it would reveal his secrets.

It is the persistence of vision time that causes rapidly changing images to blur, just as in the photograph of a runner taken with an old Brownie camera. While POV is a blessing that makes movies possible and shields us from an awareness of blinking, it also imposes limitations on our responses. A fighter throwing a punch at 30 miles per hour reaches his adversary in less than a typical POV time. Boxers have to study fighting positions and stances to glean a little extra time to parry blows. Fencing is another sport operating close to POV times. The bout between the mongoose and the cobra suggests a rapid response time for the mongoose although the creature is not always a winner.

A discussion of POV was given by the original thesaurus editor Rôget in the early nineteenth century. In 1832 the Frenchman Plateau patented a device that provided the illusion of continuous motion on a rotating disc. He also established that sixteen frames per second were optimal and that each frame had to be separated by an equal dark period for the cinematic effect to function best. The first demonstration of animated films was given in 1834 by setting up a group of lantern slides each with a slightly altered drawing. The projectionist ran past the lanterns with a light source, illuminating the pictures one at a time.

An invidious application of POV is in brief subliminal responses, a comparable effect to our insensitivity to blinking. A single frame with an advertising message is inserted into a movie and the viewer's subconscious responds to the very short message ($^1/_{30}$ sec) without his conscious awareness. This technique has been employed for commercial exploitation but it may also have beneficial effects in pedagogy. The overlap of our conscious and subconscious in times short compared to POV raises interesting questions about the way the brain stores visual information. Faces are probably stored as an array of lines, edges, curves, etc. and every new encounter is rapidly and unconsciously scanned to find a comparison. The scanning is probably repeated numerous times to identify the image. That the mechanism is not perfect is revealed by the number of times we err in believing an individual is someone else. If it is a place, it is sometimes mistaken as a form of deja vu (I think I've been here before).

Another aspect of POV is in color separation. Stare at a red image until the brain is saturated with the red signal, close your eyes and the same image, complementary in color, appears. Black spots turn white and *vice versa* when we apply this phenomenon to geometrical objects. When we read we recognize groups of letters as words rather than the individual letters since the word patterns are stored in our memory. If we only recognize a single word at a time then it means we require ten seconds to scan 100 words (0.1 sec per word). Thus, eye movement to focus on individual words becomes critical for rapid reading. Recognition of entire sentences within the POV time can increase reading speed.

POV governs much of our lives and probably restricts the pace of our activities. A driver in the Indy 500 must frequently make decisions at times close to 0.1 sec. Fortunately we tend to avoid such perilous situations but when unavoidable it can drive us to distraction.

CHAPTER XII

Early 20th Century
Part 5

Most physicists remained skeptical of Einstein's light-quantum hypothesis even after Milliken's experiments on the photoelectric effect. Arthur Compton clinched the case, however, when he showed how one can play billiards by shooting photons at electrons. By the time he performed this experiment in the early 1920's X-rays technology had advanced significantly and X rays of a single energy could be isolated.

By directing a beam of electrons at a molybdenum target, X-rays of approximately 17,000 eV of energy were copiously emitted compared to other energies. These X-rays were then aimed at a piece of carbon and collisions with electrons were recorded. Some of the X-rays were scattered back with no loss of energy while some suffered about 4% loss in energy.

One can understand no loss in energy if an incoming X-ray collides with an entire carbon atom. It would be like throwing a tennis ball against a wall - the mass of the wall is so large compared to the tennis ball that the wall would barely recoil, hence no energy loss. This follows quite simply from conservation of energy and momentum, nature's insistence that collisions of its fundamental building blocks preserve these two quantities.

The fraction of photons that underwent the 4% energy loss puzzled Compton. He speculated on a variety of wild ideas that might provide an explanation but was forced to the conclusion that if an X-ray of energy hν were considered to have a momentum hν/c, where c is the velocity of light and h is Planck's constant, that a collision with a single electron would produce about a 4% energy loss in its recoil. It meant that the photon was behaving like a particle with a precise energy and momentum, while the photon plus electron were acting like billiard balls, one small in weight and the other large.

This explanation accomplished two things - it ended Compton's frustration in looking for an explanation and it secured for him the Nobel prize, both important to a serious physicist. This experiment is one of the simplest dealing with the Creator's fundamental particles.

What was Compton like, this man who had dispersed the cloud of mystery about the particle nature of photons?

In the summer of 1942 Professor Arthur Holly Compton of the University of

Chicago was placed in charge of atomic bomb production. To implement this program, J. Robert Oppenheimer and a number of leading theoreticians met in secret at the University of California to analyze the problems associated with nuclear explosions. The fear of a major German effort drove the group, many of whom were refugees, to a strong sense of urgency. It was rumored that Heisenberg was heading a similar project in Berlin.

After a few days of analysis Edward Teller brought some dire news to the group. His calculations showed that the temperature build-up in a bomb could ignite the nitrogen in the atmosphere, in a manner similar to the sun. The entire world could blow up! With a monkey wrench this size thrown into the works, Oppenheimer had to let Compton know. However, he discovered that Compton was vacationing at a lakeside cottage in a small town in Michigan.

Reaching him on the phone in the general store, he arranged to see him in Michigan, since tight security forbade discussions by telephone. Key personnel in the Manhattan project were forbidden to fly and Oppenheimer took the tiresome train journey, examining the Teller report for errors. At a deserted spot along the lake front Compton listened to the tale of doom. Should they abandon the project? (Should Heisenberg's group be allowed to blow up the world first?) Compton hedged and asked that the group study the matter further and give him a probability for the end of the world. Back in Berkeley the group discovered a critical oversight in Teller's calculations. They set to work and produced a super-secret probability for Oppenheimer to deliver to Compton: three chances in a million for a world doomsday! The 50-year old Compton considered this news and made his decision: go ahead!

Following the discovery in 1895 of X-rays emitted by cathode- ray tubes, and of gamma rays from radium in 1900, physicists could not reconcile their properties with those of radio and light waves. But when Laue proposed the Nobel Prize experiment in 1912 of diffracting X-rays from crystals, the wave nature became more attractive than the particle. Arthur Compton discovered the effect that now bears his name and provided experimental evidence for the duality of wave and particle behavior. He demonstrated that X-rays had an energy h times the frequency and momentum h/c times the frequency (c the velocity of light). In addition the Compton effect provided a clearer picture of the electron itself.

Arthur Compton, whose older brother Karl became President of MIT, was born in Wooster, Ohio in 1892 into a family influenced by Mennonite teachings. Arthur attended elementary, high school and college in Wooster where he showed interest in astronomy, religion and gliders. He received a Ph.D. in physics from Princeton in 1916 with an interest in the quantum nature of specific heat, the subject that had concerned Einstein. Compton took his first post at the University of Minnesota, but moved into war work at Westinghouse when America entered the fight. He received some of the early patents on the sodium vapor lamp and on fluorescent lighting.

At the end of the war he came under the influence of J.J. Thomson, Rutherford and other notables during a years' visit to the Cavendish Laboratory in Cambridge. With no X-ray generator available, Compton did gamma ray scattering and absorption measurements. He unsuccessfully tried various models for the shape of the electron to account for the measured gamma-ray wavelength change after scattering. Appointed to Washington University in St. Louis, he was provided with an X-ray generator that produced a sufficiently copious source of radiation to nail down the precise mechanism. He presented his results in December 1922 at the Chicago meeting of the American Physical Society.

Compton showed that the scattering of an X-ray by an electron could be treated as a simple billiard-ball collision, with X-ray energy and momentum as predicted by Einstein and with a simple point electron of mass m. No other measurement is so fundamental to atomic physics, one photon colliding with one electron. The University of Chicago appointed him to the physics staff and he remained there for 22 years, where he measured the index of refraction of materials for X-rays (the values are less than one by a few parts in 10^5). He ruled some diffraction gratings on an engine constructed by Michelson and employed these for X-ray diffraction. In 1927 he received the Nobel Prize shared with C.T.R. Wilson.

In a two year visit to Oxford in 1934-35, he became aware of the Nazi menace and the German refugee problem, which undoubtedly influenced his later commitment to the Manhattan Project. On his return to Chicago he pursued his interest in cosmic rays and showed that the primary sources were charged particles from space deflected by the earth's magnetic field.

After Hahn and Strassmann's discovery of fission and Compton's appointment to the Manhattan Project, he arranged for the use of Stagg Field at the University of Chicago as the site for the world's first chain-reacting pile under Fermi's leadership. The early plutonium chemistry was done at the university's Metallurgical laboratory, and with Compton's encouragement du Pont undertook the construction and operation of the plutonium separation plant at Hanford in a remote part of the state of Washington on the Columbia River. He was also instrumental in setting up the uranium isotope separation facility at Oak Ridge.

When the final decision had to be made whether to drop the bomb on Japan, Compton canvassed a number of the Chicago scientists as to their recommendations and he attended a meeting called by the Secretary of War Stimson to consider their input. Fermi, Lawrence, Oppenheimer and Compton were present to hear Stimson's words,

'Gentlemen, it is our responsibility to recommend action that may turn the course of civilization. In our hands we have a weapon of unprecedented destructive power. Our great task is to bring this war to a prompt conclusion.'

All eyes turned to Compton as leader of the group but he remained tacit and it befell the articulate Oppenheimer to summarize their feelings. Compton was under great stress since Szilard had been pressuring him to recommend restraint. When someone complained to Fermi that Compton constantly referred to God, philosophy, and brotherhood, Fermi answered simply, 'That's his current need.'

Compton withdrew into his shell of childhood religious training. Having once given Oppenheimer the go-ahead on three chances in a million, should he be asked to play God twice? He died in 1962, aged three score and ten just as the Bible stated.

Five years later the Compton effect experienced a rebirth when it was realized that it was capable of the most accurate assessment of the predictions of the Schrödinger quantum mechanical description of electron behavior in atoms.

There are several stories about the precise catalysis that led to the discovery of quantum mechanics. It is unlikely that we will ever be certain because such ideas are discussed over beer, at meals, at conferences, etc. The one I find fairly believable occurred during a physics colloquium when the postulate of wave-like properties for the electron was being discussed. Someone suggested to Debye that he try to formulate a wave equation for electrons, just as Maxwell had done for electromagnetic radiation. Debye countered that he was too busy and suggested that Schrödinger try his hand. The latter succeeded with what is now recognized as the Schrödinger equation.

Earlier, Heisenberg, Born, and Jordan had approached the problem with a more complicated mathematical approach - successful, but since abandoned as too cumbersome.

In a simplistic way the Schrödinger equation for the hydrogen atom (one electron and one proton) starts with some input, notably the mass of the electron, the mass of the proton, the negative charge on the electron and the equal and opposite positive charge on the proton. It is assumed that the electrostatic force between the electron and the proton varies inversely as the square of the distance between the two, even over these submicroscopic distances. Since the mass of the proton is more than 1800 times greater than the electron one assumes the proton is at rest and the electron moves around it. This leads to a myriad of mathematical solutions but we throw most of them away - they do not agree with the real hydrogen atom. This may sound a bit like handwaving but that's the way it is - no sense in worrying about something that doesn't give the right answer. As Benjamin Franklin said, "It's convenient being a rational being since we can justify everything we do."

Each of these mathematical solutions is called a wavefunction and from it we can evaluate a number of things that can be measured. We can calculate the probability of finding the electron at various positions relative to the proton; we can calculate the probability of the electron having various values of momenta; we can calculate its total kinetic energy, its potential energy, and the sum of the two i.e. the total energy. What is the result of all this mathematical manipulation?

We discover that only those wavefunctions for which the total energy is a constant, no matter where the electron is relative to the proton, are permitted. In these cases the energy is made up of one part kinetic energy (positive) and two parts potential energy (negative). There are a limited number of such energies, hence the term quantized. If we are, indeed, rational homo sapiens we can understand that the electron will not alter its energy in its orbit about the proton, assuming there are no other bodies nearby with which to change this energy. However, the electron can switch from one energy level to another either by emitting or absorbing a photon of this energy difference. Voila! This solved the mystery of the sharp lines in the spectra of atoms - the quantized energy levels.

Suddenly things began to fit into place, as physicists began to examine the measured optical lines emitted by atoms, particularly hydrogen. There were still a few major ramifications necessary such as the realization that the electron was spinning on its axis as does the earth, introducing changes in the energy levels. But since those successful days from 1926 onwards we've never looked back. The Schrödinger equation was the key that unlocked the mystery. With Hitler's rise to power the Schrödinger story contained elements of cinematic drama.

When the Nazis marched into Austria in March 1938, the Nobel Laureate Erwin Schrödinger found himself trapped. As former Professor at the University of Berlin, he had left Germany in 1933 as a protest against Hitler's racial policies, and spent three years at Oxford before accepting the professorship at Graz, Austria. Having been a native of Vienna, he was pleased with his appointment to Graz but the *Anschluss* left him *personna non grata* with the new authorities. Although an Aryan, his outspoken rejection of the Nazis in 1933 found him unable to secure an exit visa from Austria in 1938. The Nazis kept accurate files on important people!

An unsigned message was secretly conveyed to Schrödinger, written on a 4 x 5 card, and relayed to him via his mother-in-law living in Vienna. It simply asked if Schrödinger was interested in a post at a newly created Institute for Advance Studies to be built in Dublin. The note originated with Prime Minister de Valera, himself a mathematician, and was passed on through Professor Born in Edinburgh, Professor Baer in Zurich, and delivered to Vienna by a Dutch friend of the latter. After receiving the welcome offer, the Schrödingers burned the card and, with the Gestapo breathing down their necks, they drove to Konstanz at the Swiss border where they secretly met with Professor Baer and asked him to inform de Valera of their grateful acceptance.

At the end of the summer Professor Schrödinger received a registered letter containing a three-lined message informing him of his dismissal from his post at Graz. His immediate thought was getting out of Austria but Italy was the only country accessible without a visa. The Schrödingers back-packed a few cases and left their worldly possessions behind. In three days they trekked over the mountains and arrived in Rome where they were given temporary exile in the Vatican since Erwin was a member of the Papal Academy. Schrödinger sent a note to Mr. de Valera, who was in Geneva at the time presiding over the League of Nations. The Irish prime Minister phoned Schrödinger and arranged an Irish visa to allow him and his wife to get to England. The escape from Italy is best told by Mrs. Schrödinger:

The visas took only five days and we were ready to leave Rome. But at Iselle, the last Italian station before the Simplon Tunnel to Switzerland, the police came looking for us and ordered us out of the train with our luggage. I was shown into a dull waiting room where I was bodily examined by a woman who did not speak a word. I became excited and frightened as to what might happen. After the woman finished I sat on a wooden bench and looked anxiously out the window at the train held up for us. I longed to get away. Although it was only half an hour, it seemed an eternity to me. At last I saw my husband enter the train and I was called and allowed to board.

The customs officials had become suspicious of this pair who had passports containing all the necessary visas to get to England, were travelling first class (paid for by the Irish authorities), and yet had only declared £1 each in their possessions. They must surely be smugglers! The £1 reflected the limited currency restrictions of the time.

In passing though Geneva they were met by Mr. De Valera, who urged the two to proceed to England immediately because of the worsening situation in the Sudetanland. They arrived in London just before the Munich crisis and then headed to Dublin where they remained 17 years.

Born in Vienna in 1887 Erwin Schrödinger was tutored at home and showed an early fondness for physics and mathematics. He later entered and graduated from the Gymnasium in Vienna and in 1910 he received a doctorate in theoretical physics from the University of Vienna. He stayed on as an assistant until the Great War when he served as an Austrian artillery officer. Returning to his post after the war and marrying in 1920, Schrödinger moved on to a professorship at Zurich, where he studied color-blindness in humans.

His interest in de Broglie waves and a new theory of gases by Einstein led to his famous 1926 discovery of the wave equation named after him. In terms of its usefulness in calculating and understanding optical levels in atomic systems, it is without peer. Its success in 1926 was followed by his appointment to succeed Planck in Berlin (1927). The thing that rankled the Nazis was his thumbing his nose at them and abandoning this prodigious post in Berlin a few months before he received the Nobel Prize in 1933.

During his years at Dublin he wrote a book entitled *What is Life?* in which he considered questions of spontaneous mutation.

He spent his final years in Austria (after 1950) and received many awards from the Germans probably in recognition of his shabby treatment under the Nazis.

The final key to the unravelling of atomic energy levels came with the realization that the electron was spinning like a top. The spinning electron was proposed in 1925 by Samuel Goudsmit and George Uhlenbeck, two Dutchmen who later emigrated to America. The general result of his discovery was to more than double the number of quantized energy levels. In making use of the Schrödinger equation one now had to include the quantized spin in the wave function in such a way that electrons would be indistinguishable. Here again, the rules that were adopted for deciding on proper wave functions proved to be necessary to fit the experimental data although philosophical arguments often accompanied the rules.

Goudsmit went on to play a leading role in the Allied search for the Germans working on the atomic bomb during World War II.

The 1920's and early 1930's were golden days for physicists. Most of them lived in Europe and their common devotion to unravelling the secrets of nature transcended national boundaries. Their number was sufficiently small that one was able to recognize and befriend most of the community.

Two of this group, Samuel Abraham Goudsmit and Werner Heisenberg, both outstanding theoretical physicists, became good friends. They had made fundamental contributions to the understanding of optical energy levels, a subject that has challenged the physics community from the early days of the Bohr quantized orbits for hydrogen. Heisenberg received the Nobel Prize in 1932 for his discovery of quantum mechanics, and Goudsmit came close to receiving one for discovering the electron spin.

The Dutchman Goudsmit emigrated to America in 1927 because the Netherlands could not provide employment for all its graduates. However, in spite of offers of chairs in America, Heisenberg remained in Germany. After Hitler unleashed World War II the two friends found themselves on opposite sides and, more tragically, Goudsmit's parents were exterminated in a concentration camp after the invasion of the Netherlands. With the announcement of fission to the world in January 1939, the race for the 'bomb' began. As the leading German physicist Heisenberg was placed in charge of the uranium problem, while Goudsmit spent the early war years on the radar project at MIT.

Shortly after his discovery Goudsmit was invited to lecture at the Sorbonne and he described the spinning electron as *l'electron retournant*. The audience laughed at his improvised French, and he later discovered he was describing an electron with flatulence.

His conflict with Heisenberg surfaced when Goudsmit was urgently called to Washington in early 1944. Sam had some inkling that an atomic bomb project was underway since many of his colleagues had 'disappeared' into the New Mexico desert. He was asked to head the super-secret ALSOS project and find out what the Germans were doing on the atomic bomb. Intelligence in Germany was hard to come by. The French underground could not be used since they were not privy to the existence of the Manhattan Project. The answer seemed to be to send Goudsmit into Europe after D-Day to locate the Heisenberg team since it was unlikely that the Germans would engage in such an effort without Heisenberg. At that time a successful Anglo-American bomb test in New Mexico was still more than a year away.

Heisenberg's position in Germany was ambivalent. While he recognized the potential of a nuclear explosive, he considered it to be 20 years away. Furthermore, he and Otto Hahn, the discoverer with Strassmann of fission, were disenchanted with the Nazis and were not anxious to deliver such a weapon into Hitler's hands. They did not pursue the uranium problem, as it was called, with any enthusiasm.

Shortly after the invasion of Denmark Heisenberg visited his old teacher Bohr in the hopes of arranging a secret agreement whereby the German, American and English physicists would eschew nuclear bomb work. However, Bohr was in no mood to discuss such a matter with a man from a country that had violated Denmark's boundaries.

For the most part the German scientists dragged their heels, while the American and English super-effort on the Manhattan Project reflected a fear of Hitler that overrode any calm assessment of the situation. Aerial reconnaissance over Germany failed to uncover anything remotely resembling an Oak Ridge or Hanford in magnitude. Furthermore, with overwhelming Allied air supremacy in 1944, any such discovery would have been the target for concerted bomb attacks. But in time of war one plays it with caution.

When the ALSOS mission arrived in France, Heisenberg's group had already left bomb-torn Berlin for southern Germany. The sabotage of the heavy water plant in Norway and the failure to realize that pure graphite could be employed in constructing a reactor deprived Germany of a chain reacting pile before the end of the war (they did come close).

Before Goudsmit could move into southern Germany to locate the Heisenberg team, he passed through the Hague where he found his boyhood home a bombed-out shell. His parents had saved his high school report cards and these were strewn over the floor. In spite of Heisenberg's effort to intervene with the Nazis, Goudsmit's father and blind mother had been exterminated in a concentration camp.

At the Rhine and under fire, samples of water were collected by the ALSOS group to be flown to Washington to test for radioactivity, in case the river was being used to cool an atomic reactor. Goudsmit also enclosed some scarce French wine with a humorous note to check it for activity. Washington found more radio-activity in the wine than in the water since grapes can pick up natural groundwater activity. In all seriousness, Washington demanded more wine and Goudsmit had to send one of his men on a tour of France collecting two bottles of wine from each district (one bottle for his own files).

After the ALSOS mission reached Paris, covert information was sent to Heisenberg in southern Germany to alert him that Goudsmit was looking for his group. Heisenberg was totally unaware of the Manhattan Project and assumed the Americans were after the German atomic secrets! He promptly hid his notebooks in the latrine.

By the time Goudsmit's group caught up with Heisenberg, the Gestapo were also after him to keep their 'secrets' out of Allied hands. In a shooting match Heisenberg was rescued from the Nazis, and with nine other prominent nuclear scientists (including Hahn, von Laue, Gerlach and von Weizsäcker) was taken to England for secret internment in a large unoccupied home until the Anglo-American atomic bomb was dropped. The group assumed they would be questioned about their own work, possibly asked to direct a postwar US effort in atomic energy. Unbeknownst to them and against Geneva Convention regulations, the place was 'bugged'.

While being fattened up during their several months of internment, the ten German scientists suffered from a lack of scientific journals and a decent library, one of them complaining that he had read *Alice in Wonderland* several times. The group was smug, convinced that since they had found the uranium problem so difficult, the Americans could scarcely have gotten any further. The titular head of the project, Professor Walter Gerlach, had already written such an analysis for the Nazi high command.

When news of Hiroshima was broadcast over the BBC, the Germans were unable to accept this shattering surprise - except Laue who had remained an outspoken critic of the Nazis. He was spared by the Gestapo because of his Nobel Prize and his friends in scientific circles. At first, Heisenberg considered the BBC report as pure propaganda, since his own overestimate of the size of a uranium bomb would have made it too heavy for an aircraft. As it happened the bomb fully strained the capacity of the B-29 bomber. By the time three days passed and the announcement of the Negasaki plutonium bomb was broadcast, Heisenberg had reconsidered his position and accepted the news as true. A few days of remorse followed as the group felt they had failed the Fatherland. Gerlach took it very seriously and Hahn, feeling personally responsible for the bomb, had to be restrained from harming himself. Ironically the announcement of his Nobel award in chemistry came during this period of internment. Hahn was thus doomed never to forget his innocent contribution to weaponry.

After a further period of reassessment, the Germans extricated themselves from this 'defeat' by convincing themselves they were only interested in peaceful uses of atomic energy. This was technically correct although a working reactor would have given Hitler untold quantities of radioactivity to rain on London in the V2 rockets. This would have created a panic just after D-Day and thwarted the invasion.

It was this holier-than-thou position of innocence, publicly espoused by Heisenberg after the war, that brought the two old friends into conflict. Goudsmit did not condemn Heisenberg for working on an atomic bomb - during wartime this was justified - but the secret recordings convinced Goudsmit that the Germans would have produced a weapon if it had been feasible in a wartime Germany. Heisenberg's ignorance of the secret tapes and the restraint placed on Goudsmit not to reveal their existence brought this debate into an awkward but emotional focus.

Five years after the war Goudsmit and Heisenberg were finally reconciled, yet it wasn't until 1992 that the British Foreign Office released the secret recordings. Heisenberg returned to his research in Germany and Goudsmit became editor of the *Physical Review*. Goudsmit's story is told in his book *ALSOS*, but only recently has the Heisenberg story been told.

The 1930's were relatively unproductive as both the depression and the rise of the dictators altered the lifestyles of physicists. Emigration to America accelerated but jobs were not easy to come by. There was little government support for research and not many students could afford the luxury of studying such a remote subject. The politics of science took on a new appearance as the threat of war loomed.

In the early 1930's the work of Compton was improved when DuMond at Cal Tech used a high-resolution spectrograph to examine X-rays scattered from beryllium metal. The nature of a metal is such that the outer electrons in each atom form an almost continuous band of quantized energies. This manifests itself in the distribution of the electron velocities present before the X-ray collides with the electron.

The DuMond measurements confirmed these electron velocities and provided an early corroboration of theoretical efforts to calculate these energy bands in metals.

The war years witnessed a cessation of much of the activity in optical research although some war-oriented developments like bomb sights and radar took rapid strides forward. The long wave length end of the electromagnetic spectrum from radar to radio waves has played a significant role in wartime, since transmittal and interception of information became part of the spy and underground systems. The insistence by the German high command that U boats in the Atlantic transmit weather reports to Berlin led to the defeat of the submarine threat when the wireless messages gave away their positions.

Radar owed its origin to the nineteenth-century development of radio waves, the work of Marconi dating to 1895. At this time it was suggested that there was an ionized layer surrounding the earth at a height of 60 miles, an excellent reflector for radio waves due to the strong interaction between electromagnetic radiation and charged particles - i.e. between photons and electrons. This ionosphere proved a means for bouncing radio waves around the world by multiple reflection. The height of this layer was determined in the 1920's by timing an outward and return short radio pulse. The detection of lightning flashes from the electromagnetic signal they emitted provided an aid to long-range weather pattern identification, flashes as far away as 4000 miles being picked up. In 1922 Marconi suggested the employment of short-wave radio waves to detect ships and in 1930 the use of the reflected radio waves to locate sea traffic was proposed.

Before World War II the British initiated a program of coastal reflected ray detection stations and went on in 1939 to develop the critical magnetron for producing high-intensity radar in the microwave region, 10-cm wavelength and less.

One of the ultrasecret wartime operations in 1942 was the raid on Dieppe during which the English examined and removed critical components from the German radio ranging station to determine whether they had developed anything like the magnetron. While the press reported this raid as a catastrophe for the Allies it was a resounding success in assuring the Allies that their radar lead was unchallenged. Radar provided the edge that saved Britain in 1940.

The story of the atomic bomb, the most powerful man-made source of light and blast pressure, focuses around J. Robert Oppenheimer. It is one of the tragic tales of the post-war period.

In August 1944 J. Robert Oppenheimer, director of atomic bomb construction at the super-secret Los Alamos Laboratories was faced with the prodigious problem of designing an implosion technique to ensure an efficient plutonium fission bomb. The problem was related to the spontaneous fission of plutonium (no triggering neutron needed) and the subsequent release of neutrons with the fragments. As the critical mass components are fired at each other pre-ignition may occur and produce a dud.

This did not appear to be a problem with the U^{235} bomb, but only enough separated isotope would be available in 1945 to construct a single bomb by the summer. On the other hand, plutonium production at Hanford would soon be in full swing.

The successful D-Day landings in 1944 added a sense of urgency. If the war ended before the atomic bomb had been used the entire *modus vivendi* at Los Alamos would be dissipated. Operating with no budget restraints, Oppenheimer conceived the idea for Jumbo, a 25-foot diameter steel chamber into which the test bomb would be placed. Any pre-ignition would keep the valuable plutonium from being strewn over the desert.

This 5,000-ton chamber could only be fabricated on the East Coast; and under the extreme security regulations of the Manhattan Project the massive container was transported on a special multi-wheeled vehicle over roads without tunnels or bridges, late at night, and away from the eyes of the curiosity seekers. By the time this expensive operation had been completed at the end of the year, adequate experimental data was available to give Oppenheimer sufficient confidence to dispense with the need for Jumbo. After that it sat in the desert for 20 years before being dismantled at a cost far exceeding the original!

The atomic bomb days at Los Alamos from 1943 to 1945 were a far cry from the 1920's when the young Oppenheimer was at Göttingen working on the early problems in quantum mechanics. Not many associate Oppenheimer with the field of optics, but the well-known Born-Oppenheimer approximation is so widely employed in calculating atomic energy levels that one hardly gives it a second thought. The problem taken up with his teacher Max Born was to apply the Schrödinger equation to molecules and solids. Can one account for the vibrations of the atoms without the undue mathematical difficulty in following both electrons and nuclei? By recognizing that the lighter electrons would always follow the nuclear movement closely, Born and Oppenheimer suggested the nucleus be assumed stationary while the electron behavior was calculated, and then repeat this for other positions of the nucleus, finally averaging the electron distributions over the nuclear positions. In 1944 such considerations were child's play compared to calculating the instantaneous changes in a plutonium bomb as it first imploded and then exploded.

Oppenheimer came from a well-to-do family. He never worried about money and he followed his own spirit as he matured. With his natural brilliance he studied physics at Harvard under Bridgman, Nobel Prize winner for his work on high pressure, and graduated with a B.S. in 1925. In those days a sojourn in Europe was *de rigeur* and Robert went off to the Cavendish Laboratory in Cambridge. Physics consumed him with a passion and a sense of frustration - the subject almost drove him to suicide. He consulted an English psychiatrist, who diagnosed *dementia praecox* and told him not to return, he was too much of a responsibility. Oppenheimer went off to Corsica, where he fell in love with the countryside and returned cured. He later met Dirac who was puzzled that Oppenheimer would spend time reading Dante. No physicist idled away at such trivia - perhaps the violin or piano but definitely not Dante!

Born granted Oppenheimer a Ph.D., in spite of his having forgotten to register at the university when he first arrived. The 'Wunderkind' was impetuously ill-mannered and would go to the blackboard in the middle of the Professor's lecture to describe a better way to derive an equation. Student complaints forced Born to hint to Oppenheimer that such behavior was unacceptable. Born and Oppenheimer did not cross paths after that. Returning to New York, Oppenheimer decided to drive to California but, lost en route in thoughts about physics, he drove up the courthouse steps in one town and came to a halt at its massive doorway.

He lectured alternately at Cal Tech and Berkeley but, feeling the need of a refresher course, he returned to Europe to study with Pauli in Zurich. This only lasted a few months and one can only guess that these strong personalities could not co-exist in the same city. Excessive work has its toll and Oppenheimer developed tuberculosis, moving to the desert air of New Mexico to recuperate. Enchanted with the mountain scenery, it probably accounts for his subsequent decision to recommend New Mexico for the bomb assembly program.

In September 1929 he was back at Berkeley teaching the exciting new quantum mechanics. Students had difficulty following his physics lectures but they were mesmerized by his style. When unable to follow his scientific reasoning they focused attention on his facility in manipulating chalk and cigarette simultaneously, certain he would either write with the cigarette or put the chalk in his mouth. As time passed the lectures became more lucid and he developed his own American school of aficionados. His genius at grasping scientific essentials and translating them into appropriate language was without peer. His voice and demeanor entranced his audience and commanded respect from those who came under his spell. Unfortunately, he suffered no fools and left the more ignorant permanently wounded by his caustic barbs. This became his Achilles Heel!

Students would emulate him. He remarked that he did his best work from 2 to 5 AM and one student sneaked into his home for an early morning 3 AM session only to find him asleep! Oppie read no newspapers or magazines and didn't listen to the radio. He learned of the stock market crash six months after the event. During the Depression that followed, many of his students were drawn into Communist causes. Some, like his physicist brother Frank, joined the Party. A girl friend introduced him to the Spanish Loyalist Cause and for several years Oppie devoted time to fund raising. He was courted by left wing faculty members and subscribed to the Communist People's World. All this would later come back to haunt him during the McCarthy purge of the 1950's.

He married in December of 1940 and was soon drawn into the excitement of the discovery of fission. When Compton assumed command of the secret atomic bomb development program, the most likely man to head the theoretical effort was G. Breit of Yale, but a personality conflict led Compton to choose Oppie as a substitute.

A room in the attic of the physics building at Berkeley was set aside for their deliberations. In the absence of cogent experimental data the entire bomb technology had to be analyzed theoretically. As a member of the group Edward Teller estimated that an atomic bomb reached temperatures sufficiently high to detonate the atmospheric nitrogen and turn the earth into a fiery sun. Oppenheimer immediately contacted Compton who was on holiday in Michigan. The war with

Germany was already underway and the German refugees in the theoretical group viewed their mission as one to beat Heisenberg to the bomb.

Oppenheimer journeyed to Michigan to meet with Compton, a man of Mennonite upbringing, and asked him for guidance. No doubt Compton thought of the Biblical Great Flood. Could a Noah survive a fireball? He asked Oppenheimer to have the group place a probability on such a disaster. On his return to Berkeley the group discovered an oversight in the Teller calculations and were able to report that the probability for the world to end as a result of the detonation of an atomic bomb was three chances in a million. Compton decided to give it the green light and wondered if Heisenberg had been presented with similar concerns.

As the Manhattan Project proceeded into high gear, General Groves, in overall command, needed to appoint someone to assume charge of the design and assembly of the bomb. He spoke to a number of candidates but only felt comfortable with Oppenheimer. Groves was not going to place the program in charge of some 'screwball' scientist who spoke in mathematical unintelligibility, whereas Oppie could reduce complicated ideas into simple concepts. Groves sustained considerable criticism from the Army security branch who dug up all of Oppie's pre-war left wing activities, including his Communist brother. Anxious to get on with the job, Groves rejected the adverse reports. Army security never stopped their probing and managed to get other scientists close to Oppie fired. Many physicists were surprised at the choice, particularly those who had a personal score to settle, such as Teller who was passed over for selection as head of the theory group. Unable to get Teller to apply himself, Oppie relieved him of his duties and permitted him to do theoretical work on the hydrogen bomb, of low priority during the war.

Following Oppenheimers' appointment in November 1942, he and Groves selected a remote site in New Mexico to ensure that the bomb's design group could work in isolation. Critics of Oppie felt that Groves was one of many mesmerized by his unnatural sway over people. While Los Alamos was under construction, Oppenheimer set out on a recruiting campaign, Very few of America's leading physicists could resist the persuasive charms of the man who couldn't even tell them why they were being asked to leave their university existence for the isolation of the desert. Yet he was successful, and one wonders who else could have accomplished the task.

Once at Los Alamos there was only one way out for its residents, either divorce in the case of a wife or by quitting for good. Petty problems such as an unusual high birth rate, inadequate diaper supplies, limited entertainment facilities, dust and more dust made life exasperating, and Oppie's role formidable. He became the last authority on all matters, yet he seemed always to produce the right answers. whether domestic or scientific. Most people have described his leadership as 'magnificent'.

When the first samples of plutonium pointed to a large probability for spontaneous fission and the release of fast neutrons, it became obvious that the U^{235} and plutonium bombs required different approaches. The uranium bomb consisted of two subcritical masses fired at each other to detonate the rapid chain reaction.

The pieces had to coexist long enough to ensure that all of the uranium underwent fission. The bomb, ultimately released over Hiroshima, would have been the preferred type if U^{235} had been available in large quantities. The need to use plutonium threw Oppie's life into a nightmare since the only way to force the subcritical pieces together long enough to ensure high efficiency was by implosion, a technique new to ordnance experts. Developing a new technology, the pressure to succeed before the Germans did or before the war ended, and the internal problems at Los Alamos caused Oppie to lose considerable weight and take on a cadaverous appearance.

At the end of 1944 the ALSOS mission under Sam Goudsmit had uncovered enough information in Europe to show that Germany was not a threat. It is ironic that this did not alter the determination of the group who would have gladly stuck with Oppie to complete the job even if the war had ended. Pure scientific curiosity would have provided the momentum.

In order to establish the size of the critical mass a series of experiments called 'Tickling the Dragon's Tail' was undertaken in which a subcritical piece of fissionable material was permitted to drop through a hole in a second piece of subcritical material so that, for a fraction of a second, the combination was sufficiently close to criticality to produce a large multiplication of emitted neutrons. On one occasion the physicist L. Slotin became careless and the reaction proceeded sufficiently long to produce a lethal dose. He became one of fission's first victims. Oppenheimer would often be a bystander to this mesmerizing experiment.

Following the successful test of the plutonium bomb in July 1945, the burden of responsibility momentarily fell from Oppie's shoulders, only to be resumed when he was called in by Truman for his opinion on the bomb's use. He took a *cautious stance* based on military considerations, not realizing that the decision to drop the bomb may have been partly political - to *keep Stalin out of the final peace negotiations with Japan*.

Physicists like to brag about their accomplishments. They do so by publishing in journals, by showing visitors around their laboratories to look at their latest experimental achievements, and by presenting papers at scientific meetings. The bomb detonation in the New Mexico desert was the biggest showcase of all time but it was impossible to turn the clock around. Even with Germany gone and Japan on her last legs it would have been contrary to expectations to deprive those associated with the Manhattan Project of their test shot.

Hiroshima catapulted Oppie to world fame, just as the Eddington experiment on the gravitational attraction of light rocketed Einstein into prime press copy. For the latter, money appeared from lecture fees and voluntary contributions to his causes. For Oppie, it came from the largest single source in the world, the US defense budget. Senators and Congressmen now realized that physicists could perform military miracles and one took note of their utterances! Add to this the hypnotic demeanor of this genius and you have a target comparable to the President.

Oppenheimer's elevation to chairman of the powerful General Advisory Committee, and the successful detonation of a Soviet bomb in 1949, years before anticipated, created the ingredients for a modern tragedy. Oppie had offended several key people who had uttered ignorances in science; the fear of the Russians had produced a McCarthy; and the race for the hydrogen bomb had become a symbol for national survival. Oppie was too powerful not to be a sitting duck. His difficulties started when the General Advisory Committee turned down a proposal to develop the hydrogen bomb since it was deemed not technically feasible. Later, the committee reversed itself when a new solution was found, but at the time personal antagonism prompted some to urge removal of Oppenheimer as a stumbling block to the country's survival.

Every piece of ancient left wing history was raised to discredit Oppie, and Eisenhower revoked his security clearance, thus removing him from his country's services in atomic weapons. He could easily have been asked to quietly resign, but when he was publicly charged for past sins that were already known during the war, he chose to fight. The tenor of the times, as with Galileo, prevented a fair hearing. That trial has been effectively portrayed in a BBC series.

Though subsequent efforts were made under President Kennedy to acknowledge Oppie's service with the prestigious Fermi Award, it was too little too late. A few years later he died, still a security risk.

CHAPTER XIII

Latter Half of the 20th Century
Part 1

Optical development in the second half of the twentieth century took gargantuan leaps with the semiconductor, the laser, the optical fiber, magnetic resonance, and astronomical observations exploring the Big Bang. One might add that light also penetrated the USSR when it decided to westernize its economy.

Until its semiconducting properties and its ability to store information were discovered at Bell Labs about 1950 the element silicon was known to be abundant but almost useless in its pure state. Its development for information storage mushroomed when Brattain, Bardeen, and Shockley demonstrated that by adding special elements localized junctions were formed that could store an electrical pulse. The advantage to the semiconductor was the ability to rapidly read, write, and erase this information and to pack a large density of these junctions on the surface of a single crystal wafer.

While the semiconductor field was being developed for computers the first successful laser was demonstrated by Ted Maiman in 1960 employing a single crystal of ruby i.e. sapphire (Al_2O_3) grown with a small percentage of Cr atoms replacing the Al atoms. Sapphire is ordinarily colorless but the addition of the chromium atoms imparts a red color due to its three quantized energy levels. The lowest energy or ground state can absorb light energy and be raised to a higher energy quantized state, followed almost immediately to an intermediate energy state by the emission of 'invisible' infrared light. This intermediate state is metastable i.e. it will hang around for a while before dropping down to the ground state by emitting a red photon. However, it is possible to coax it to return it to its ground state almost instantaneously by passing a red photon of the identical energy within a reasonable distance of the chromium atom. This is Einstein's process known as stimulated emission. Once stimulated emission occurs you have two red photons, the one used to tickle the chromium atom and the other emitted when the chromium returns to it ground state. These two photons can stimulate two more Cr atoms producing four red photons, then eight, etc. - a chain reaction creating identical twins with each birth.

The important thing about stimulated emission is that the proton you stimulate is a clone - identical in energy, direction, polarization, and phase to the proton used to provide the stimulation. As this process continues until all the intermediate states in the ruby have returned to the ground state, you rapidly create a concentrated beam of photons all heading in the same direction, hence the term laser i.e. *light amplification by stimulated emission of radiation.* In order to enhance the power output the ends of the ruby are half silvered to reflect half of the red photons back into the crystal and stimulate any chromium atoms missed the first time around.

The metastable intermediate energy level of the chromium atoms in ruby comes about because the electron has a very different orbit in the two states and it must wait for the right set of conditions in order for the electron to move from one quantized state to the other. The photon that provides the stimulated emission acts as a guide to show the way.

The requirements for any such laser are an intermediate metastable state on a substance that can be activated or 'pumped up' by an external photon source. Since the work of Maiman in 1960 the optical engineers have found all varieties of gases, solids, and liquids that possess these properties so that lasers covering a wide spectrum of photon energies and materials are commercially available.

Once the laser made its debut engineers addressed the problem of directing the light beam to its target. Mirrors and lenses are well-known possibilities and are often part of laser systems, but it was the optical fiber that proved most versatile as a waveguide (light conductor). By purifying silica (SiO_2) and drawing it hot into fine fibers of diameter less than 200 microns, it was established that at certain wavelengths photons would travel for kilometers without noticeable loss in intensities, provided there we no bends in the fiber. A simple improvement, though, of adding a thin cladding material with index of refraction about 1% less than pure silica enabled fibers to undergo nominal bends without loss i.e. the light bounced back into the fiber. Only at very large angles of bend did the light escape through the cladding.

CHAPTER XIV

Tutorial 5
Photons Versus Electrons

Until the commercialization of optical fibers, electricity was the prime method for transmitting signals and information between points, but a comparison (1994) of the two techniques, light and electricity, reveals clear overall advantages for the optical fiber. Some of the points of comparison are:

1. Low resistance - Both electrical signals and optical signals can be conveyed for miles without excessive loss, both at close to the speed of light. In this respect the two techniques are comparable.
2. Polarization - Optical signals can be distinguished by their polarization unlike electrons which can not be polarized in a wire. The optical polarization give the photons added versatility.
3. Coherence - The coherence of the laser beam launched through an optical fiber permits several beams to be separated at the detector according to their phase. Electrons are never coherent, at least no one has devised a technique for creating and launching such an electrical signal.
4. Two-dimensional Flow - Laser beams can be simultaneously launched and detected at both ends of a fiber without interference - not possible for electrons with can only flow one way at a time.
5. No Electromagnetic Interference - Lightning flashes will wreak havoc with electrical signals, not so with photons traversing an optical fiber.
6. Broad Band - Optical fibers can transmit a broad range of wavelengths at the same time, enhancing the information content. Electricity can only transmit one signal at a time.
7. Vibration Sensitive - Optical fibers are more sensitive to stresses along the fiber, not true with electrical signals, except for specially designed pressure transducers. This is one area of advantage for the electrons.
8. Short-circuit hazard - While electric lines are subject to short circuit fire hazards, optical short circuits are generally harmless to the environment, unless one is transmitting high power beams for cutting.
9. Compatible with Silicon Chips - Since silicon chips are based on electrical signals within the chip the electrical wire is generally compatible for

connection to the chip. Optical fibers demand some special kind of interface although efforts are underway to directly read chips optically by directing a pinpoint of laser light from above onto the junctions. Development along these lines will enable one to increase the information density in a chip since the electrical connectors simulate a railroad switching yard - it gets difficult to add more track.

10. Superconductivity - A superconducting optical fiber has not been discovered although a vacuum is equivalent since light will pass through it unhindered. However, such a hollow core fiber is not generally viable since bends in the fiber will cause the photons to crash into the wall and suffer the probability of being absorbed. As for electrical signals the room temperature super-conductor has eluded us but, if it were found, electrons may resume some advantages over photons.

11. Non-linearity - In certain materials photons can excite the atoms to a higher energy state but before the atom returns to its ground state other photons passing through the material will experience a different index of refraction. The potential for non-linearity, i.e. the more initial photons striking the sample the greater is the change in the index of retraction, is under intense exploration. In such materials the index of refraction depends on the optical power passing through the material. This has not been pursued for electrical conductors although it can not be ruled out i.e. a strong electric current can change the state of a sample, hence the resistivity. The whole point of employing non-linearity is to make optical fibers sensitive to optical power, a further example of their versatility.

More information per unit weight of optical fiber can be transmitted compared to copper wire. The two techniques are still competitive when costs are considered and I'd need a crystal ball to predict the technology ten years hence.

Optical fibers are now extensively employed in medicine, not only for diagnostic testing within the body but for therapeutic delivery of lasers to remove or destroy unwanted material. Between optical fiber technology and magnetic resonance scanning (radio waves) the electromagnetic spectrum continues to be exploited for man's benefit.

CHAPTER XV

Latter Half of the 20th Century
Part 2

The magnetic resonance phenomenon was independently discovered by Bloch at Stanford and Purcell at Harvard in the 1940's, both receiving the Nobel Prize in 1952. The technique relies on the properties of certain atomic nuclei that possess spin like the electron, the most important nucleus being the proton itself. When a substance containing hydrogen is placed in a magnetic field the magnetic moment of the proton takes on two quantized energy states, which can be viewed as the magnetic moment being parallel or anti-parallel to the magnetic field direction. Under these conditions the protons in the sample collectively precess around the magnetic field, like a spinning top, but when photons whose energy equals the difference in energy between the quantized states enter the sample they elevate the hydrogen nuclei from the ground state to the excited state. With appropriate detectors this photon absorption can be measured even though the quantized energies are small, typically at radiofrequencies.

For many years after its discovery magnetic resonance primarily interested physicists but more recently the medical profession has discovered it as a non-invasive method to 'see' inside the body. The patient is positioned within a large magnet with a magnetic field strength about ten thousand times stronger than the earth's. The hydrogen atoms in different parts of the body possess slightly different quantized energy levels so that the technique can concentrate on specific parts of the body, depending both on the placement of the detector and the precise radiofrequency employed. The technique is called nuclear magnetic resonance by the physicists but the doctors have dropped the word nuclear so as not to alarm patients. It has the advantage over X-rays in producing images of parts of the body without causing damage to cells.

In the 1950's a totally unexpected response to photons was discovered by Mössbauer in Germany when he found atomic nuclei that absorbed X-rays. Prior to this work one only considered electrons as the means for scattering and absorbing photons, any direct interaction with nuclei was considered too small to be measurable. The unusual effect is named after the discoverer who received the Nobel Prize in 1961.

It was first isolated in the Fe^{57} nucleus which has an excited state 14.4 keV above the ground state. The energy difference between the ground and excited state is very precise - one part in 10^{11} so that is requires a photon of exactly the right energy to be absorbed by the Fe^{57} nucleus. If the photon has been Doppler shifted by moving the target as little as 1 cm/sec the effect disappears! Pound at Harvard made use of this by showing that a photon that fell in the gravitational field of the earth gained the expected energy in terms of its Doppler shift. The 1919 Eddington measurement of the gravitational attraction of starlight by the sun had catapulted Einstein to world fame. Pound showed that the Mössbauer effect could accomplish the same result in the laboratory. The Mössbauer effect is limited to but a handful of nuclei and within a few years of its discovery physicists had exhausted their ideas in making use of it and it is now rather passé.

Man's serious study of light began with observations of the visible photons entering our atmosphere from the sun and the stars. Five hundred years after Leonardo's work astrophysicists have again assumed an exciting role in man's quest for knowledge by examining the very origin of all existence, the Big Bang. We earthlings occupy an infinitesimally small portion of the universe yet we seen to be able to accumulate information and develop theories that may enable us to attempt to rewrite Genesis.

It started in the 1960's when the subject of radio astronomy i.e. searching the skies for electromagnetic radiation in the radiofrequency range became fashionable. The signals at these frequencies were found to totally fill the skies with the distribution of frequencies corresponding to a black body with a temperature within a few degrees of absolute zero (2.75 K). This is now interpreted as the residue of a Big Bang dating back billions of years. As everything blew outward after the Big Bang the gases slowly cooled to the 2.75 degree temperature, except for regions like the stars where matter condensed gravitationally and nuclear reactions kept them at high temperatures.

Almost fifty years before this discovery Einstein had considered his General Theory of Relativity but his picture of the heavens puzzled him. If the stars were stationary then the universe should be unstable, gravity would cause it all to collapse. Not eager to predict a collapsing doomsday he was forced to include in his theory an unidentified cosmological force of repulsion to overcome gravitational attraction. In the 1920's Hubble's observations of the Doppler red shift indicated that the visible universe was not stable but was moving outward in all directions just as if it had exploded from a point in space. The relative speeds were measured at about 30km/sec for two galaxies that were a million light years apart. Working back in time indicated that the universe started from a Big Bang some 15 billion years ago.

If the entire universe all began at a single point we must be careful not to assume that those physical quantities we now identify as time, energy, mass, and charge would have had an identical meaning to those we share at present. But assuming these quantities are meaningful at some time after the Big Bang we can build up certain ideas about its details. The cosmic black body background discovered in 1964 by the 1978 Nobel Prize winners Penzias and Wilson is consistent with the cooling down of an expanding gas. These photons (radio waves) are quite uniform throughout the sky and indicate a present temperature of 2.75 degrees above absolute zero. This background may represent the equilibrium conditions of the first hundred thousand years after the Big Bang.

Theoreticians have been able to calculate such quantities as the ratio of the abundance of the light elements and these agree with the observation that there is 10^8 times more helium in the universe than lithium. However, unanswered are such questions as the observed 10^{10} ratio of photons to protons in space and why the universe is very homogeneous when averaged over distances of hundreds of light years.

Of course, the history of physics has shown that ideas change, sometimes quite rapidly. Right now the Big Bang is fashionable and an entire school of study has attracted adherents but it may become old hat by the time you finish this book.

Optical engineering, however, appears to be growing faster than the Big Bang. Scientists are examining every conceivable method for utilizing optics in conventional engineering such as:

1. Entertainment; A successful use of multicolored laser beams can be seen at several planetaria in the US and abroad. Employing a high-powered gas laser with a helium-argon mixture to produce four colors, the display of laser lighting is projected onto a planetarium ceiling with mirrors. Music fills the theater as the laser lights change in rhythm to the beat and create aesthetically exciting images, some in animation. The music serves as the story line with the laser acting to embellish the flow of tunes and tempos. This may be only a passing fad, at least until the laser animation begins to provide the narrative.

    Small lasers are commercially available to enhance lecture presentations by such techniques as using a dancing company logo to follow specific items on a slide or viewgraph.

    Compact discs (CD's) employ a simple diode laser to read digitized musical data. Since the data are permanently etched into the disc their lifetime is quite high before deterioration sets in. Unlike magnetic tape, CD's are permanent and cannot be erased but that is being worked on.

2. Dentistry: Patient's face dental drilling with trepidation. The vibrations transferred to the teeth and bones are no one's cup of tea. Even high speed diamond drills transmit frequencies that shatter one's equilibrium. Along comes the laser that does not contact the tooth but merely evaporates the caries - no unwanted vibrations. But there are problems associated with the laser such as the lungs ingesting the evaporated material, particularly the mercury from old fillings. Dentists are reluctant to invest in such expensive equipment until the technique becomes commercially competitive and free from lawsuits. In the interim companies are expending R & D funds to break into the market.

3. Healing of burns; Quite accidentally it was discovered that red lasers

directed against skin burns accelerated the healing process, particularly when treatment followed soon after the accident. It is a guess that the laser is somehow accelerating the natural healing process by which the immune system recognizes and discards dead cells.

4. Robotics; Robots are not a generic species that look and behave like humans. They can be unpretentious devices that merely replace man in the performance of simple manufacturing processes. They can execute welding operations with continuous information feedback to assess whether the weld is properly executed. More importantly, robots can be useful in performing under hazardous conditions such as in the presence of toxic gases, explosive danger, high radiation levels etc. The construction and training of robots is currently a fashionable area of research and the possibilities appear unlimited.

5. Medicine; The laser-bearing optical fiber endescope carries both laser power and visual signals enabling the surgeon to destroy unwanted tissue by evaporation or cell destruction. Port wine stain, a reddish discoloration of the skin made famous by Gorbachev, has been a cosmetic triumph of the profession. But more serious disorders such as blockages in arteries have been corrected. The optical fiber permits the surgeon to view the exact spot he is irradiating as well as enabling him to closely follow the results of his treatment. It can frequently be performed without the risks that accompany major surgery, a local anesthetic may be sufficient.

6. Optometry; The sculpting of the cornea to correct visual faults can eliminate the need for glasses. The removal of cataracts and relieving glaucoma pressure by drilling a hole in the cornea can be accomplished painlessly in a single office visit.

7. Optical sensors; It is possible to introduce small sensing media in an optical fiber so that the optical signal will register temperature, pressure, electrical voltage and current in a wire, specific gas content in the air, ph, and even geographical position with a fiber gyroscope. In the last item two laser signals from the same source are launched in opposite directions around a fiber loop. An acceleration of the loop will produce a different Doppler shift in the opposed light beams and this can be read interferometrically and tabulated to keep track of position. When these devices are cheap enough they will be installed in autos and will minimize the need for maps by registering one's exact position. Sounds like sci-fi but it's on our doorstep.

8. Radioactivity; Fibers imbedded with a scintillating substance will yield an optical output upon exposure to radiation.

9. Magnetic Fields; A fiber imbedded with magnetite will alter its shape in the presence of an external magnetic field. This will affect an optical signal traversing the fiber and enable one to calibrate the signal relative to the strength of the magnetic field.

10. Bodily Functions; Fibers are so thin that they can be inserted harmlessly

into the body just as an acupuncture needle. These fibers can register a number of in-vivo functions such as ph, blood flow, chemistry, voltage etc. One can perform one's ordinary chores while the fiber continuously monitors the vital signs.

11. Mechanical Integrity; Bridge, airplane, and building supports can be continuously monitored under the strains of external stresses by inserting optical fibers into its structural members.

12. Home Lighting; A single powerful source of light in the basement can provide illumination to various parts of a house through fiber conductors.

13. Holography; Holograms have emerged as a consequence of the coherent properties of the laser i.e. each photon emerges in phase with the others. When a laser beam is reflected from an odd-shaped object and is then recorded on a photographic plate it becomes a phase picture of the object. When reconstructed it provides a clear, lifelike picture. Holography is a rapidly growing field that is gaining acceptance as an artistic technique. It is difficult to predict where it is going.

14. Liquid Crystal Displays: Liquid crystals consist of long chain molecules the size of visible light that selectively diffract depending on the wavelength and the molecular size. Certain liquid crystals change their molecular size with temperature and are employed as simple oral thermometers, changing color with temperature.

The above list will steadily increase over the years for it is difficult to think of an area of technology where optical fibers and laser sources are not likely to be employed. Optical computing is emerging as a new discipline with predictions to provide orders of magnitude increased capacity for data.

The five centuries from 1500 until the present have been witness to an experimental growth in optical physics and engineering. The highlights of this period include the genius of the great painters and cinematographers, the theories of Newton, Maxwell, and Einstein, and the development of the laser, the optical fiber, the transistor, and the computer.

But the rapidity with which new discoveries emerge leaves one at a loss as to where to place your bets on future developments. The race to discover new ways to use photons and atoms has become an exciting game and it often drives society faster than it can cope. Optics appears the key to the future storage and retrieval of data but filling our libraries with trivia does not necessarily answer the question of our mission in the Creator's scheme of things.

SUMMARY

Light as a method for communicating aesthetic appreciation through art, architecture, cinema etc. has created a group of mortals almost diametrically opposed to the approach of the physicists. The former are keen on unlocking the techniques for transferring emotion through visual stimulation while the latter must approach their experiments with the detachment of hard experimental data uninfluenced by personal desires. One rarely sees individuals professionally steeped in both disciplines - the personality requirements for such antithetical approaches leaves the two groups incommunicado.

Physicists will certainly visit art galleries and the cinema, but they rarely develop creative urges in these fields. Artists are probably interested in learning some physics but the subject is difficult and off-putting. 'Tis a pity universities don't offer degrees in artistic physics. As we bring this story to a close we might have one last bird's eye view of the curvy road that mankind has followed in exploring and exploiting his most valuable stimulus.

At the beginning of this evolutionary road we cannot fail to note that sleeping is the only biological function that encourages one to bypass visual stimulation (except for dreams). In all other of our experiences sight is the stimulus most relied upon although a loud noise or a hot stove will promote responses that bypass the urgency of the visual. I knew someone blind from birth until a teenager. After restoration of his sight it took a long time for him to trust his visual rather than acoustic stimulation.

The road forks in the 17th century with the signpost identifying the wave theory path and the corpuscular lane. We have forgotten which path we took as the road mounts a hill where an observatory stands and a sign makes all welcome to stay the night, observe the sky, and listen to a lecture by Sir Isaac Newton. His talk is unintelligible but soporific so that we fall into a deep slumber. We awake in the 18th century and take note of Dr. Franklin wearing his bifocals. We can see that the two roads are heading in the same general direction as we traverse the 19th century, finally converging at King's College London were Professor Maxwell lectures us about the unity of electricity and magnetism. Back to sleep.

We sail into the 20th century with Einstein at the helm but are again put to sleep listening to him speak of relativity. He lights his pipe and takes out his violin - that arouses us. We moor at the pier only to witness Professor Oppenheimer pointing to a huge mushroom cloud. We discover it is but the echo of the Big Bang. As darkness descends a policeman lights the way with a red laser emerging from an optical fiber. Red means danger and I ask whether I can take a trip back to enjoy some vino with Leonardo. I'm informed that time does not run backward but who knows what's on the other side of the velocity of light?

So what's ahead? Man is beginning to move afield of his homeland and will soon (in the scheme of things) populate new horizons in space at speeds that may be too slow for him. If man solves the genome question he could transmit his genetic code at the speed of light to a far-off solar system where he could be reincarnated. But in his infinite ignorance man has developed an arrogance that overshadows that of Count Rumford. The nature of things is so complex that man can not formulate sensible questions about his insignificant role in the Great Drama. Basically, he has to realize that he can be no greater than himself.

Benjamin Franklin, one of the Creator's success stories, confessed that if it were possible to be preserved in a cask of brandy, to be revitalized in two hundred years, he would gladly submit to the experiment. Some souls have expressed the desire for similar treatment by immersion in liquid nitrogen. As for me, I would prefer to be beamed into space on a coherent laser beam with full instructions on genetic assembly. Unfortunately, it would be just my luck to have the instructions misread and I'd be sent back as defective. On second thought, though, even if the instructions were read properly the recipient might still rightly claim the product to be defective.

There are particles called neutrinos, much more elusive than photons. The evidence for their existence is on the borderline of believability since we know of nothing that can readily grab the critters as easily as electrons grab photons. Perhaps the Creator keeps the whole show going with even more slippery buggers than neutrinos. Whatever the case may be these days, the science pages of our newspapers frequently rival scandal and politics for our attention.

But no matter how complex science appears to be when we read the Scientific American we must remain thankful for little things, like photons and electrons. A day without sunshine is like a day without the Creator.

SOURCE BOOKS AND FURTHER READING

1. THOMAS YOUNG, *Alexander Wood Oldham*
   (Cambridge University Press 1954)
2. JAMES CLERK MAXWELL, *Ivan Tolsky*
   (Canongate Press)
3. WILLIAM HENRY FOX TALBOT, *Hutchinson Benham*
   (London 1977)
4. ROBERT HOOKE, *Margaret Espinass*
   (Heineman 1956)
5. SIR CHARLES WHEATSTONE, *Brian Bowers*
   (IEE, London)
6. LEWIS CARROLL, *Edward Guiliano*
   (Crown 1982)
7. EINSTEIN, *Ronald W. Clark*, Avon
   (New York 1973)
8. BENJAMIN FRANKLIN, *Carl van Doren*
   (Viking, New York 1973)
9. MONET, *Charles Stuckey*
   (Park Lane, New York 1985)
10. NIELS BOHR, *Ruth Moore*
    (Alfred Knopf, New York 1966)
11. EDISON, *Rex Beasley*
    (Chilton, New York 1964)
12. THE UNKNOWN LEONARDO, *Ladislao Reti*
    (McGraw Hill, New York 1974)
13. MAX PLANCK, *Scientific Autobiography*
    (Greenwood 1968)
14. GALILEO, *E.E. Levinger*
    (Julian Messner, New York 1952)
15. ALSOS, *S.A. Goudsmit*
    (Henry Shuman, New York 1947) (Goudsmit)
16. HEISENBERG'S WAR, *Thomas Powers*
    (Alfred Knopf, New York 1993)
17. DOCTOR WOOD, *William Seabrook*
    (Harcourt Brace, New York 1941)
18. THE LIFE AND TIMES OF REMBRANDT, *J. Van Loon*
    (Bantam, New York 1957)
19. MARIE CURIE, *Robert Reid*
    (Mentor, New York 1975)
20. LAWRENCE AND OPPENHEIMER, *N. Davis*
    (Da Capo, New York 1986)

176

21. BENJAMIN THOMSON, COUNT RUMFORD, *Sanborn Brown*
    (MIT Press 1979)
22. MR. GILLRAY, *Draper Hill*
    (Phaidon, London 1965)
23 MICHELSON, ALBERT ABRAHAM, *Bernard Jaffe*
    (Doubleday Garden City 1960)
24. A.A. MICHELSON, *Bennett*
    (McAllister and Cabe, App Optics, OSA 1973)
25. DE VALERA, SCHRÖDINGER AND THE DUBLIN INSTITUTE,
    Sir Wm. McRea (1970)
26. HOW TO TELL THE BIRDS ETC., *R.W. Wood*,
    (Dodd Mead, New York 1917)
27. GRIFFITH, *Martin Williams*
    (Oxford U 1980)
28. ENCYCLOPEDIA BRITTANICA

Additional Suggestions:
Barbara Cline, *The Questioners*
    (Reprinted University Chicago Press)
Alan E. Beyerchen, *Scientists Under Hitler*
    (Yale University Press)
Einstein & Leopold Infeld, *The Evolution of Physics*
    (NY: Simon & Schuster 1938)
Philipp Frank, *Einstein, His Life and Times*
    (London: Jonathan Cape 1948)
George Gamow, *Thirty Years That Shook Physics*
    (Doubleday 1966)
Banesh Hoffman, *Albert Einstein, Creator and Rebel*
    (NY: Viking 1972)
Dorothy Michelson Livingston, *The Mastery of Light;*
A Biography of A.A. Michelson
    (NY: Schribner's, 1973)
Walter Moore, *Schrödinger Life and Thought*
    (Cambridge University Press, 1989)
Rosalynd Pflaum, *Grand Obsession; Madame Curie and Her World*
    (NY: Doubleday, 1987)
Emilo Segrè, *From Falling Bodies to Radio Waves* (NY: Freeman, 1984)
    From X-Rays to Quarks (SF: Freeman, 1980)
Einstein and Leopold Infeld, *The Evolution of Physics*
    (NY: Simon and Schuster 1938)
Susan Quinn, *Madame Curie*
    (Simon and Schuster, 1995)

MP 851-2 Tn
22